To Mike

From JET : Sullom Voe, 9/81

"gig of a lifetime — thanks Mike"

SHETLAND'S OIL ERA

Special Edition

ISBN 0 904562 12 3

© Shetland Islands Council 1981

Published by the Research and Development Department of the Shetland Islands Council,
93 St. Olaf Street, Lerwick

Director: John M. Burgess

Printed by The Shetland Times Limited, Prince Alfred Street, Lerwick, Shetland

Foreword

In less than a decade the United Kingdom's oil production has grown to an extent where it more or less equates with consumption. We now rank as a major oil producer, amongst the top ten in the world league table. This is a remarkable achievement by any standards, and one of which we can be justifiably proud. The Sullom Voe oil terminal is a vital part of our offshore oil infrastructure, and plays an important role in the continuing success story of North Sea oil.

Not so many years ago, the thought that oilfields, traditionally associated with hot and distant lands, could exist beneath the cold unfriendly waters of the North Sea, would have seemed almost bizarre. Who then, standing at remote Sullom Voe, could have predicted the construction of this terminal, which is, and will remain central to the realisation of the nation's strong energy potential? Twin pipelines now converge from afar, arteries carrying the life blood, oil, on which industrialised society today depends.

The building of the terminal would not have been possible without a high degree of co-operation between the various parties involved. Their interests have not always coincided. Nevertheless, all concerned, notably the oil companies, the construction industry, the Shetland Islands Council and the Government have successfully combined their efforts to create this symbol of British technological achievement, which, when fully operational, will be amongst the largest terminals of its kind in the world. As Sullom Voe handles and processes increasing quantities of oil, it will add to the well-being of the Shetland Islands and the UK as a whole.

This book recounts much of the drama and excitement which accompanied the discovery of oil, the development of the oilfields and the construction of the Sullom Voe terminal. I am sure that you who read it will, as I did, find it a moving and inspiring account of the first decade of Shetland's oil era.

The Rt. Hon. David Howell, M.P.
Secretary of State for Energy.

Acknowledgements

As author of this publication, I wish to thank everyone who has in any way helped in its creation. Special thanks are due for:

The Foreword: to the Right Honourable David Howell M.P., Secretary of State for Energy, for his generosity in providing the book with a fine beginning;

Reading the Script: to Ian R. Clark, Executive Member, British National Oil Corporation, whose encouragement and interest has encompassed all three books in this series;

Providing Information and Statistics: to all those companies and organisations whose story is told here, especially the following:

Airwork Limited, Air Ecosse, The Bank of Scotland, BNOC, Bolts, BP Petroleum Development Limited, British Gas Corporation, Bristow Helicopter Group Limited, British Airways, British Airways Helicopters, British International Telecoms, British Telecoms, The Civil Aviation Authority, Chevron, The Clydesdale Bank, Conoco, Dan-Air, Department of Energy, European Investment Bank, Grampian Regional Council, Hay & Company, Highland and Islands Development Board, Institute of Petroleum, J. M. Ironside, Lerwick Harbour Trust, Malakoff Limited, Mobil, the mud companies herein mentioned, Norscot Oil Services Ltd., North Scottish Helicopters, Ocean Inchcape Limited, P&O Ferries, the Police, the Royal Air Force, especially Shetland Radar, Royal Bank of Scotland, Schlumberger Inland Services Inc., Scottish Council for Development and Industry, Scottish Development Agency, J. & M. Shearer, Shell UK Exploration and Production Limited, Shetland Line, Shet-Link, Shetland Health Board, UKOOA, Unionoil, Wimpey Marine Limited.

Art Work: to Helen Dickson and Brian Gamble.

Explanations and/or Checking: to all those who so patiently explained "how it works" particularly and who helped with detailed charts, Mr Robert Bruce of Sandlodge, Stewart McDonald of CeBo, Bill Semple of the CAA, Alan Butler of British International Telecoms, Daphne Duffy of OBIS, Michael Morgan of the Offshore Petroleum Exploration Service, J. ter Heide of Shell, Fl.Lt. David Dryden of Shetland Radar, John Ogden of Wood MacKenzie & Co. Ltd., Sam Steadman, Director of Consumer Protection, Dumfries and Galloway Regional Council.

Photographs: are acknowledged as printed, but many thanks are due to Dennis Coutts and BP Pet. Dev. Ltd., for their help and advice.

Typing: to those lightsome Shetland ladies Marna Williamson and Penelope Smith who worked so hard and so cheerfully to get this ready on time!

Elizabeth Marshall,
Principal Assistant,
Research,
Shetland Islands Council.

February 1981.

List of Contents

Foreword

Acknowledgements

Introduction

Chapter One	The Discoveries of Oil and Gas Offshore Shetland	11
Chapter Two	Planning for Oil	25
Chapter Three	Marine Supply Services, Supply Boats and Supply Base Developments	31
Chapter Four	Air Services	69
Chapter Five	The Sullom Voe Terminal	92
Chapter Six	The Port of Sullom Voe	126
Chapter Seven	Telecommunications	137
Chapter Eight	Oil Related Infrastructure and Social Considerations	143
Chapter Nine	The Economic Implications for Shetland of the Oil Era and Its Associated Developments	159
Chapter Ten	Summary	174
	Appendix	175
	Glossary	234
	Sources of Further Information	240

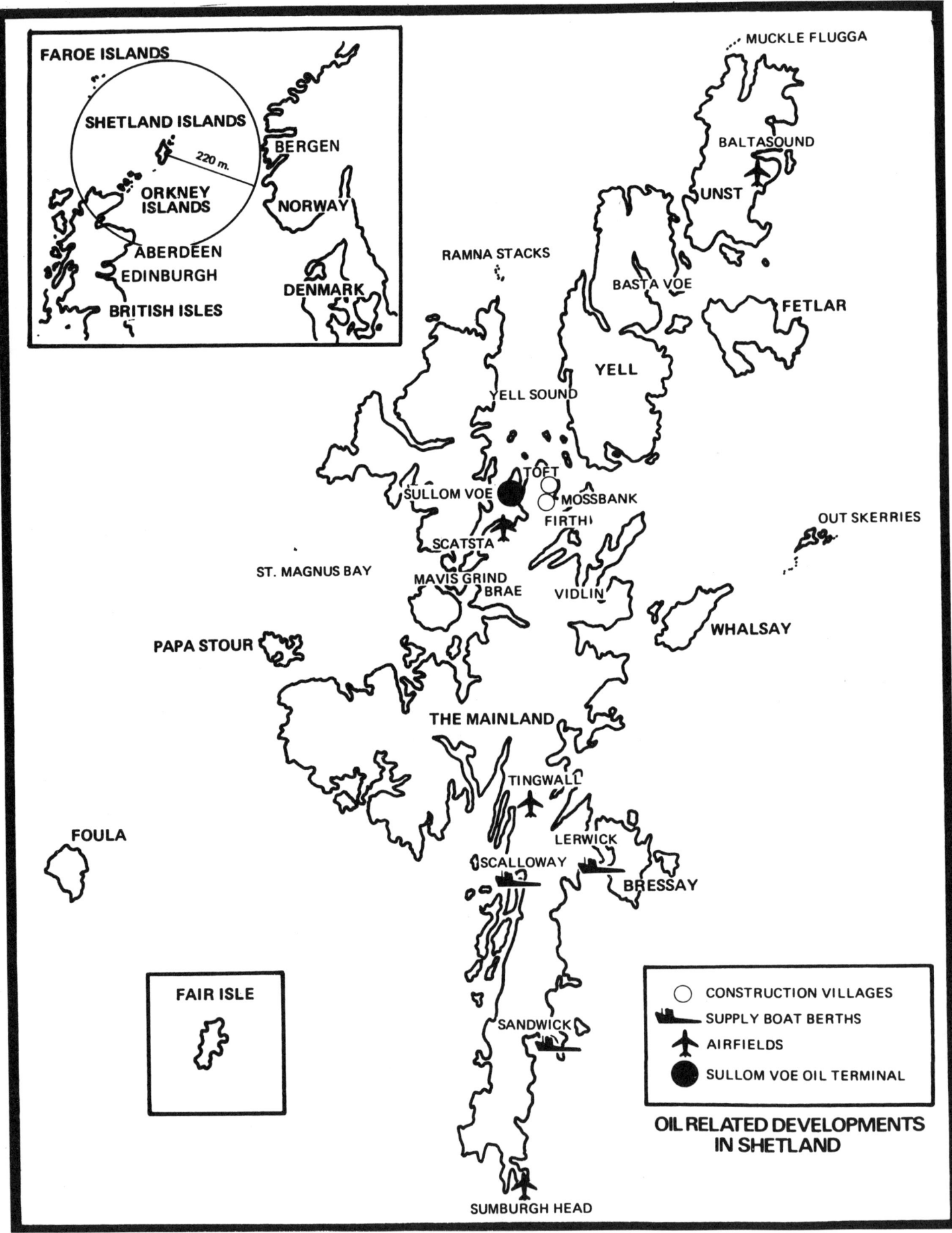

Shetland's Oil Era

Facts are not infrequently stranger than fiction. This book is a book based on facts. These facts relate, in the first instance, to how a remotely situated group of islands, known as the Shetland Islands or simply Shetland, became involved with the international oil industry. Secondly, the consequences of this involvement are examined both in terms of those developments that have occurred to date, and those which are projected for the future.

Few people realise that Britain's farthest outpost — Shetland — is located at a latitude of 60 degrees North, on the same parallel as Alaska, nor is it generally realised that Shetland is as far from London as the South of France.

Shetland is considered remote not only because it is situated at so northerly a latitude but also because the islands themselves lie in the North Sea, equidistant from the mainland of Scotland to the South, and Norway to the East (see map). This means a journey of 180 miles or more across an often inhospitable North Sea is needed* in order to reach either of the two nearest major ports of Aberdeen or Bergen.

In 1971, the 16 inhabited islands of the 100 or more in the Shetland Group contained a close knit, family conscious, rural community of approximately 17,000 people. They found their livelihood within their own community or in seas surrounding the islands. Fishing, agriculture and knitwear were and still are the cornerstone of Shetland industry. Since then, great changes have occurred, resulting from the discoveries of reserves of oil and gas beneath the same seas which were the major support of the islands economy.

Shetland is now at what might be described as the centre, of one of the world's offshore oil provinces. (See Table I). A new era has emerged in the long history of these islands — Shetland's Oil Era.

It is not easy to give a precise date as to when it began; some might consider that the birth of Shetland's Oil Era was heralded by the Shell/Esso announcement of the discovery of one of the largest oil and gas fields in the North Sea — The Brent Field — in August 1972.** Regardless of when it actually began it cannot be doubted that by the time the first oil from offshore was piped onshore to Sullom Voe in Shetland, in November 1978,*** Shetland's Oil Era had arrived.

See Graph A — North Sea Conditions compared to other similar Areas.
**The Brent Field was actually discovered in June/July 1971 but Shell/Esso made no announcement of this find until 13 months later, for both commercial and technical reasons.*
***First oil ashore was on 25th November, 1978, from the Dunlin Field via the Brent pipeline system.*

THE WORLD'S MAJOR OIL PROVINCES* (as at 1st January, 1981)　　　　　　　　　　Table I

	Estimated Proved Reserves (000 BBLS)	Country	Estimated 1980 Oil Production (000) B/D
1.	165,000,000	SAUDI ARABIA	9620
2.	64,900,000	KUWAIT	1400
3.	63,000,000	U.S.S.R.	12050
4.	57,500,000	IRAN	1280
5.	44,000,000	MEXICO	1960
6.	30,000,000	IRAQ	2600
7.	29,000,000	ABU DHABI	1380
8.	26,400,000	UNITED STATES	8650
9.	23,000,000	LIBYA	1780
10.	20,500,000	CHINA	2170
11.	16,700,000	NIGERIA	2100
12.	14,800,000	UNITED KINGDOM	1600

COMMENT: The above information shows the United Kingdom Continental Shelf (U.K.C.S.) to be twelfth in rank worldwide in terms of estimated proven oil reserves and the U.K. the ninth largest oil producing nation. As noted in other charts, in terms of both oil production and reserves the Shetland area is expected to account for more than 50% of the U.K. Continental Shelf resource for the next decade or beyond.

*Source: Oil & Gas Journal. December 29th, 1980.

GRAPH A

COMPARISON OF OPERATING CONDITIONS IN VARIOUS AREAS

▨	Maximum height wave in feet (water depth of 500')
☰	% of time where supply operations are suspended
◨	% of time where drilling operations are suspended

SANTA BARBARA CHANNEL
- Max wave height: ~40 ft
- Supply suspended: NIL
- Drilling suspended: NIL

GULF OF MEXICO
- Max wave height: ~70 ft
- Supply suspended: ~5%
- Drilling suspended: NIL

EAST COAST CANADA
- Max wave height: ~100 ft
- Supply suspended: ~10%
- Drilling suspended: ~5%

GULF OF ALASKA
- Max wave height: ~100 ft
- Supply suspended: ~20%
- Drilling suspended: ~15%

NORTHERN NORTH SEA
- Max wave height: ~100 ft
- Supply suspended: ~25%
- Drilling suspended: ~15%

Scale: 0 — 20 — 40 — 60 — 80 — 100 ft/%

Source of Information: Shell UK Exploration and Production

One of the first oil rigs into Breiwick Bay, Lerwick. *(Photograph courtesy of Dennis Coutts)*

The 433 feet long Semi-submersible Pipe Laying Barge Semac 1, in March 1977 when it began laying the Shell/Esso FLAGS pipeline. *(Photograph courtesy of Shell)*

The 85,000 ton tanker 'Esso Warwickshire' moored to Brent 'Spar' floating storage and tanker loading buoy in 1976. Both 'Esso Warwickshire' and Shell's 72,000 ton 'Drupa' were specially modified to load crude oil from 'Spar'. On the left can be seen Brent 'B' production platform.

Aerial view of Shell/Esso's Brent 'C' production platform, March 1980. In the foreground is the accommodation rig 'Treasure Hunter' housing operatives engaged in the hook-up on Brent 'C'. *(Photographs courtesy of Shell)*

Chapter One

The Discoveries of Oil and Gas Offshore Shetland

BACKGROUND

From the mid 1960's rapid growth occurred in the pace of exploration for oil and gas in the more southerly part of the UK North Sea. Exploration for oil and gas offshore Shetland was initiated in the Second Round of UK North Sea licensing, of November, 1965,* when the Department of Energy allocated some licences to companies wishing to explore in the area. However, attention continued to be focussed further south, not least because of the high rate of discoveries of oil and gas per number of wells drilled. Further, the waters of the Southern North Sea are relatively shallow, thus exploration and drilling costs are lower compared to operating in the deeper, stormier waters of the Northern North Sea (see Graph A). It was necessary for a new technology to be developed to cope with the more severe environment of the north and to develop the potential of the resources North of 59 degrees North.**

It was not until after the Third Round of licensing in 1970 that Shetland became involved in the pattern of activity.

In 1971, exploration offshore Shetland began in earnest. The strategic position of the islands as the nearest landfall for offshore vessels operating in the UK North Sea, North of latitude 59 degrees North, was utilised (see map). Supply and helicopter bases were established in Shetland. These facilities could have been of a temporary small scale nature had the exploration for oil and gas been unsuccessful. Successful exploration, exemplified by the discovery of the Brent Field, confirmed that oil and gas was present in sufficient quantities to be successfully produced. This has resulted in a prolonged period of intense offshore activity with consequent onshore developments in Shetland (see map).

However, onshore developments were to entail far more than the establishment of forward air and sea bases to support offshore activity. The oil and gas discovered offshore in the UK North Sea Sector has by law to be transported to and landed in the UK.*** The nearest practicable point at which this could be achieved for the Brent Field was Shetland. Consequently, a major crude oil loading terminal with attendant facilities to service the Brent Field and subsequent discoveries in the area was planned.

The First Round of UK North Sea licensing was announced in 1964, and licence blocks allocated in 1965. No blocks in waters around Shetland were allocated in this Round. In the Second Round of 1965, Third Round of 1970/71 and all subsequent Rounds, blocks offshore Shetland have been allocated. For further explanation see Appendix.

**For the severe conditions of the Northern North Sea, a new breed of floating semi submersible vessels, new structures such as concrete platforms, heavy lift and pipelaying barges, and new deep diving technology vessels have been designed. In addition considerable changes in the logistics of transportation have occurred, both in respect of supply boats and helicopters.*

***The Petroleum and Submarine Pipelines Act 1975, forms part of the UK legislation which insists on UK Sector North Sea hydrocarbons (oil and gas) being landed in the UK.*

The further exploration which has ensued since the Brent discovery in the East Shetland Basin* and to the West of Shetland** established this area as one of the world's major offshore sources of oil and gas. As stated, onshore facilities needed for coping with this situation have been developed. However, it is to the offshore situation that we must first turn, in order to understand the consequent onshore developments.

Developments offshore Shetland can be regarded as:

1. Developments up to the present time;

2. Future Developments (after 1981).

1 Offshore Developments up to the Present Time

The extent and rapidity of the oil and gas discoveries which have been made offshore Shetland, is indicated in the accompanying charts. These charts are divided into three categories:

(i) Named Fields for which development Plans have been announced (see Chart I in Appendix to Chapter One);

(ii) Named Fields for which there are no development Plans yet announced (see Chart II in Appendix to Chapter One)

(iii) Unnamed discoveries which may yet be developed (see Chart III in Appendix to Chapter One)

Such is the pace of exploration and development that within a short time of this book being published these charts will probably be outdated. Some Fields will have development plans announced, new discoveries will be made, some Fields may have their estimated recoverable reserves reassessed. The charts are nevertheless significant in that they give an indication of the scale of offshore discoveries which have affected, and will continue to affect Shetland.

The extent of these resources can be assessed if they are considered within the context of the national UK resource. In the first instance, one can look to the position of the oilfields offshore Shetland as part of a total of 26 producing oilfields in the UK North Sea (see graph B). From graph B shown, it can be seen that over 50 per cent of these UK resources of oil are to be found beneath the seas around Shetland. The graph shows these resources in volumes of recoverable reserves of oil, measured in barrels. This can be viewed in context of 1981 UK needs of approximately 1.8 million barrels per day of oil.*** Further illustration of the position can be gained by viewing the development of the oilfields from a financial aspect. Up to late 1980, it has been estimated that the total UK North Sea expenditure in both offshore developments and onshore facilities required to support these developments was in the region of £20,000 million (1980 money). Over the next 15 years including the remaining cost of the first 26 fields now in production, money needed for development could be as much as another £40,000 million.****

The East Shetland Basin is a geological area to the North East of Shetland where most of the important oil and gas discoveries have occurred to date.

**Significant oil finds West of Shetland were first made by BP on Block 206/8 in July 1977, named the Clair Field in August 1980, and expected to be developed towards the end of the 1980's — (See Chart II — Named Fields with no Development Plans yet announced) and Chart B in Appendix to Chapter One.*

***1.8 million barrels per day are equivalent to about one tenth of the total daily requirements of Europe. It had been estimated (November 1980) by a British National Oil Corporation spokesman that there are sufficient UK reserves to allow a production of two million barrels per day until the year 2000.*

****See John Raisman, Chairman Shell UK Limited "Oil and Gas — more to come from Scottish waters?" Given to 'Energy in the 90's' Conference, Aviemore. September 1980.*

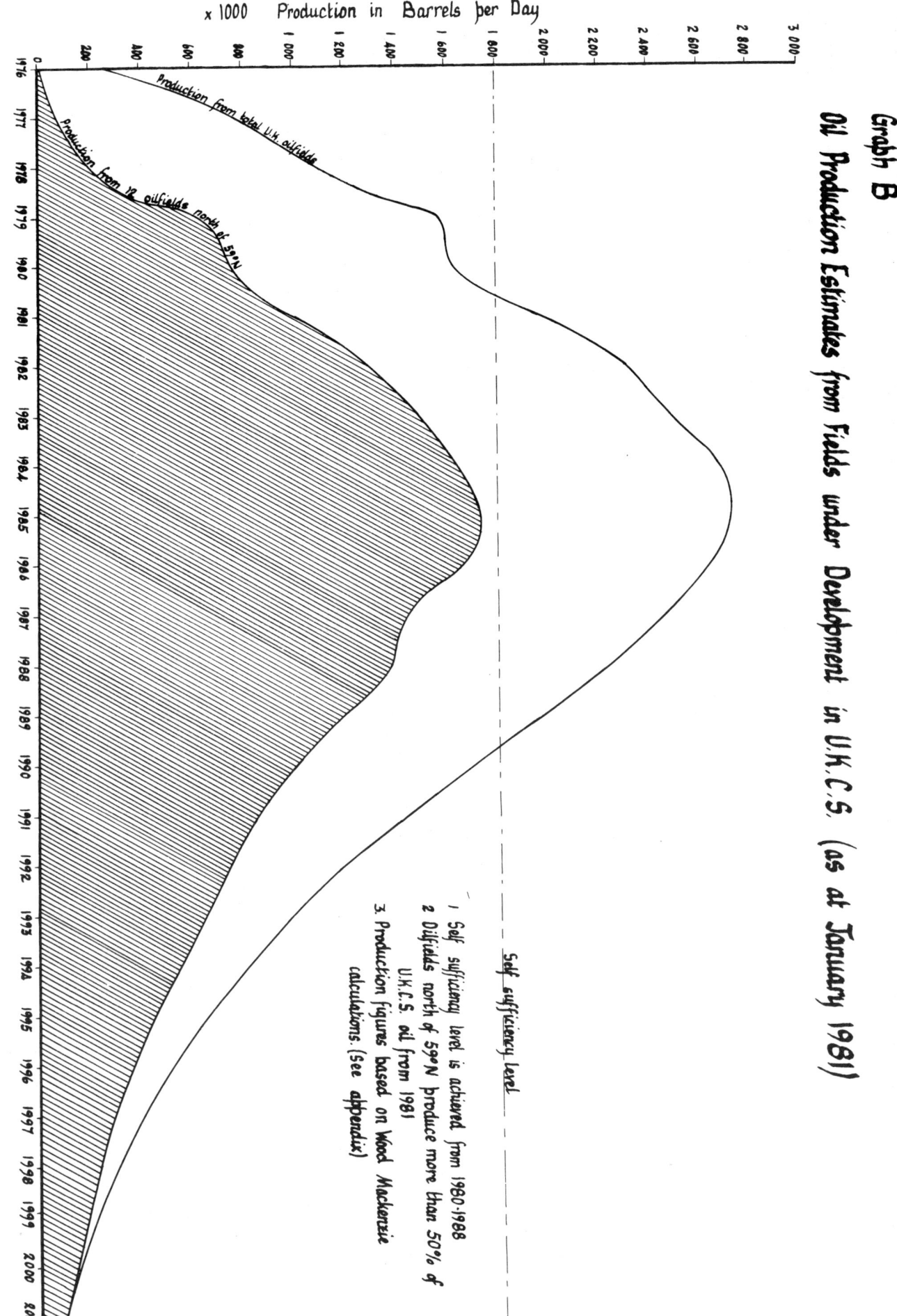

THE INCOME FROM UK OIL AND GAS FIELDS UNDER DEVELOPMENT NORTH OF 59° N (IN $ MILLION) — AS AT JANUARY 1981

Chart IV

Fields	Revenues	Royalties	Supp. Tax	Operating Costs	PRT	Depreciation	Corp Tax	Capex	Net Flow
BERYL	50683	5673	6894	6120	12083	2403	8604	3330	7980
BRENT	166984	18598	23853	19598	31376	8147	32921	9400	31200
CORMORANT	41199	4373	5939	6928	7557	3531	5774	4950	5752
DUNLIN	12975	1364	1071	1792	1627	1511	2684	1810	2627
FRIGG (UK)	19819	2137		2104		1637	7046	1822	6710
HEATHER	4961	479	99	1135	158	760	1108	840	1142
HUTTON	17424	1872	1931	2489	3726	1400	2836	1950	2619
N W HUTTON	19822	2203	2355	2347	5243	1206	3122	1550	3032
MAGNUS	37894	3885	4932	7080	3597	3126	6746	5400	6253
MURCHISON	20858	2284	2152	2850	5617	1005	3436	1223	3297
NINIAN	60174	6320	8732	10732	7612	4536	10773	5600	10416
STATFJORD (UK)	65149	7199	6211	10128	23110	1227	8297	2729	7740
THISTLE	24655	2656	2908	3633	4481	1731	4325	2460	4193
TOTALS	542597	59043	67077	76936	106187	32220	97672	43064	92961
TOTAL UKCS FIGURES	829563.28	88819.22	96949.3	126899.22	160990.34	57015.28	143985.24	79013.62	133472.78
% OF UKCS NORTH OF 59° N	65	66	69	61	66	56	68	54	70

Total UK Government take from fields under development north of 59° at Jan. 1981 = $329979 million
Total Government take from fields under development on entire UKCS at Jan. 1981 = $490,744 million
Total UK Government take north of 59° N = 67%
These figures are estimates taken over the life of the fields concerned.

FOOTNOTES:
NOTE 1   Price of oil is taken as $37.50 per barrel
2   Total UK Government Take = Royalties + Supplementary Tax brought in from 1.1.81 + Petroleum Revenue Tax + Corporation Tax. For Details of these taxes see Appendix
3   Southern UK Gas Fields are not included in these figures
4   Calculations based on £1 = US $2.38

Figures courtesy of Wood MacKenzie & Co.

ABBREVIATIONS USED:
Supp. Tax = Supplementary Tax
PRT = Petroleum Revenue Tax
Capex = Capital Expenditure
Net Flow = Net Cash Flow
UKCS = United Kingdom Continental Shelf
Corp Tax = Corporation Tax

A substantial percentage — some would estimate as high as forty per cent — of this money has been and will be spent to develop the resources offshore Shetland.

An example of the financing needed for a selected number of producing fields offshore Shetland is given here to show the magnitude of the costs involved. (See Chart IV "The Income from UK Oil and Gas Fields under Development North of 59° N")

The investment or Capital expenditure required to develop these resources, and the Gross Revenue, or income from this investment, is estimated. The income is then divided between the UK Government in the form of Royalties, Petroleum Revenue Tax, and Corporation Tax* and the oil companies concerned. The net cash flow gives one an indication of the net cash income generated by the oilfields. The sums of money involved are of such magnitude as to be nearly incomprehensible to the ordinary person. It is not surprising therefore that the total financing of these projects is beyond the scope of the National UK money markets and is conducted on an international basis, utilising the world "money pool" and Banks from locations as far apart as Belgium and Texas.

2 FUTURE DEVELOPMENTS OFFSHORE SHETLAND

One point that is made quite clear in Chart I and is further shown in the graph of 26 UK oilfields presently producing,** is that the oilfields only retain maximum output for a short time before reserves decline — in other words, the resources are limited. One is therefore faced with the crucial question "What of the future?"

Further developments are required to maintain UK self sufficiency in oil and gas, and will depend upon two major factors:

(a) How much oil and gas remains to be developed;

(b) The pace, or rate, at which this development can occur. These factors are critical to any attempt to determine how long Shetland's Oil Era might last.

(a) The Oil and Gas Remaining to be Developed Offshore Shetland

In the ten years or so since serious exploration and exploitation of oil and gas resources has been conducted offshore Shetland, one could estimate that perhaps half the reserves in place may be committed to be developed (see Chart I). Even if this estimate is accurate, the pattern of development of the remaining reserves will not necessarily be the same as in the past. Future developments will not just be of those accumulations of oil and gas already discovered (see Charts II and III)

A substantial increase in exploration is felt to be essential by the oil industry, if the UK desires to maintain self sufficiency beyond the late 1980's. A trebling of the present rate of exploration drilling, is estimated to be necessary, as it is believed that the most easily accessible, and perhaps the largest, reservoirs have already been identified. Further exploration and exploitation will occur in a more hostile and hence costly environment. Stronger ocean currents, greater water depth, and more complicated geological structures are factors expected to be in evidence as the search for oil extends to the North and West of Shetland.

The rising prices of oil and gas*** reflects the increasing awareness world wide of the finite nature of the resource. It also probably means that ultimately most discoveries made to date will be developed.

In 1976 only the largest of fields — perhaps those in excess of 400 million barrels of recoverable reserves — were thought commercially viable. Changing circumstances, both in terms of an unprecedented rise in oil prices and new technologies will probably mean most of the discoveries made to date (as seen in Charts II and III) will be exploited by the 1990's.****

---

*See Appendix to Chapter One for further explanation.
**See Graph B
***See Graph C "The Rise in the Price of Oil 1976-81". A $50 barrel or greater is anticipated by some for 1985.
****See Table II based on Grampian Regional Council Department of Physical Planning Report, September 1980.

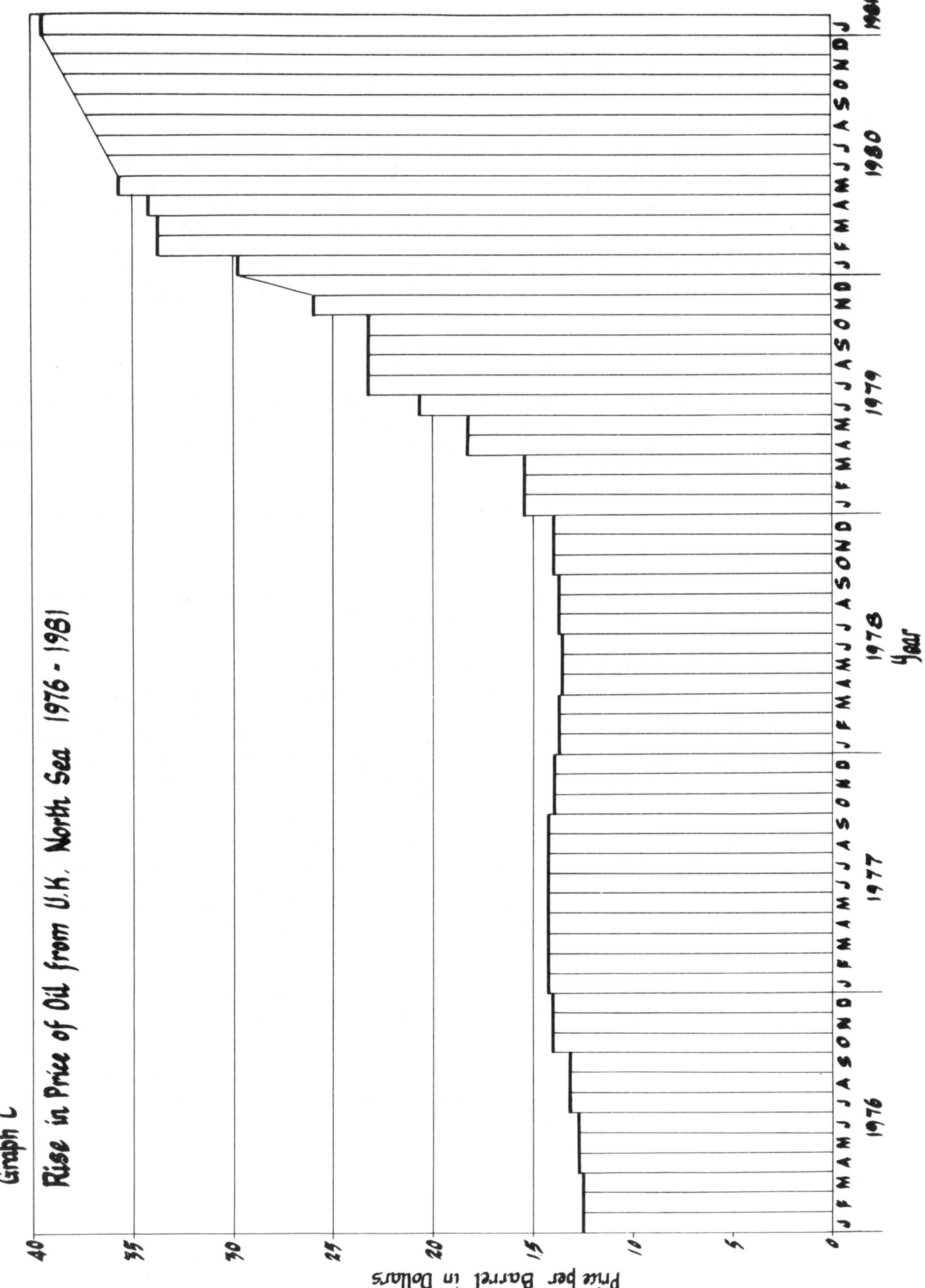

CHRONOLOGICAL ANALYSIS OF DEVELOPMENTS OFFSHORE SHETLAND Table II

(Source: Grampian Regional Council, Dept. of Physical Planning, September 1980)

Developments are placed in three categories:
1. Producing Fields
2. Fields with Development Plans Approved
3. Likely Future Field Developments

	Field	Operator	Start Up	Recoverable Reserves (million bbls)
1. Producing Fields	Beryl	Mobil	1976	400
	Brent	Shell	1976	2215
	Frigg	Total/Elf	1977	Gas
	Thistle	BNOC	1978	500
	Dunlin	Shell	1978	400
	Heather	Union	1978	150
	Ninian	Chevron	1978	1200
	Cormorant (South)	Shell	1979	110
	Murchison	Conoco	1980	380
2. Fields with Plans Approved	Cormorant (North)	Shell	1982	400
	N.W. Hutton	Amoco	1982	280
	Magnus	BP	1983	480
	Beryl B.	Mobil	1984	300
	Hutton	Conoco	1984	260
3. Likely Future Developments	Tern	Shell	1985	
	3/7	Chevron	1986	
	Alwyn	Total	1986	
	Eider	Shell	1987	
	North Thistle	BNOC	1987	
	2/10	Chevron	1988	
	Bruce	Hamilton	1988	
	9/18	Conoco	1989	
	211/13	Shell	1989	
	9/19	Conoco	1989	
	Clair	BP	1990	

Within the next five years offshore in the waters east of Shetland many new developments are expected when the UK Gas gathering pipeline* for the Northern North Sea is eventually built. Even if there are delays to the present programme and the pipeline is commissioned later than the projected date of October 1984, this will not alter the fact that the development of a gas gathering system will allow many fields to be developed that might otherwise not have been, due to the difficulties of stabilisation and transportation of gas. Apart from oil fields with associated gas (noted in graph B) in the Northern Sector of the UK system, namely Magnus, Murchison, Thistle, NE Thistle and UK Statfjord and some in the Central area such as the Beryl complex which are already in production, many other smaller accumulations might become viable.

(b) **The Pace or Rate at which Future Developments are expected to occur Offshore Shetland.**

Future developments offshore Shetland referred to here will concern:

(i) **Fields already in production or under development (see Chart I).**
All fields in this category were discovered before the end of 1975. This is significant because the previous Labour Government in UK gave assurances (known as the "Varley Assurances")** after the Government Minister concerned), which was reaffirmed by the present (1981) Conservative Government, that there would be no production restraints before 1982 on fields discovered before the end of 1975.

GAS FLARED FROM FIELDS OFFSHORE SHETLAND IN 1980 Table III

Field	Average per day (in million of cubic metres)
Beryl	0.38
Brent	3.45
South Cormorant	0.41
Dunlin	0.61
Heather	0.10
Murchison (UK)	0.51
Ninian	1.51
Statfjord (UK)	0.20
Thistle	0.61
Total average gas flared daily off Shetland	7.78 million cubic metres

COMMENT:

In 1980 gas flared offshore Shetland was approximately 65% of the UK total daily average, with replacement cost (i.e. cost of lost resource) estimated at approximately £250 million.

In 1980 the gas wasted was about half the amount of gas wasted during the peak period of flaring in summer 1979. In order to avoid waste from flaring, companies are now taking steps to reinject gas into fields, where possible, to be produced later when a £2 billion gas gathering pipeline network is in operation.

(Figures based on Department of Energy estimates)

---

*See Appendix to Chapter One for background to gas gathering in the Northern North Sea.
**See Appendix to Chapter One for listing of Varley assurances.

Restraints on flaring gas can affect the production levels in oilfields, until the fields have facilities installed to handle gas — either to separate it and transport it by pipeline or to reinject it into the structure. The UK Government introduced restraints on flaring in the last quarter of 1979. This meant that some Fields such as the Brent Field, discovered before 1975, had to cut back oil production temporarily until they had installed gas separation and treatment units. The Government's objective was not however, to cut production from the oilfields but simply to avoid wasting the valuable gas (see Table III "Gas Flared From Fields Offshore Shetland")

OIL PIPELINE STATUS OFFSHORE SHETLAND AS AT DECEMBER 1980   Table IV

Pipeline (from . . . to)	Operator	Length (Miles)	Diameter (inches)	Status (when laid)	Terminal Status (if applicable)
NW Hutton to Cormorant A	Amoco	8.5	20	1981	
Thistle — Dunlin	BNOC	7	16	Complete	
Magnus — Ninian Central	BP	54.7	24	1981	
Ninian — Sullom Voe	BP	103	36	Complete*	Under construction*
Hutton — NW Hutton	Conoco	4	20	1982/3	
Murchison — Dunlin	Conoco	10	16	1980	
Beryl B — Beryl A	Mobil	5	20	1983	
Brent System (Cormorant A) to Sullom Voe	Shell Expro	96	36	Complete	Under construction*
Dunlin — Cormorant	Shell Expro	21.5	24	Complete	
North Cormorant — Cormorant	Shell Expro	10.5	20	1980	
Eider — Cormorant	Shell Expro			Not yet available	
Heather — Ninian Central	Union	20.5	16	Complete*	

COMMENT:

Total oil pipeline laid or to be laid offshore Shetland = 294.2 miles
(i)   16 inch =   17 miles (completed)
(ii) 20 inch =   23.5 miles
(iii) 24 inch =   54.7 miles
(iv) 36 inch =   199 miles

   This represents an expenditure probably in excess of £290 million

FOOTNOTES:
*Throughput in December 1980 at the Sullom Voe Terminal from Ninian and Brent system pipelines was in excess of 700,000 barrels per day on average, with construction 87% complete.

Source of information: Daphne E. Duffy, Petroleum Review, January, 1981.

GAS PIPELINE STATUS OFFSHORE SHETLAND AS AT DECEMBER 1980

Table V

Pipeline (from . . . to)	Operator	Length (miles)	Diameter (inches)	Status (when laid)	Terminal Status (if applicable)
N.W. Hutton — WELGAS line	Amoco	8	10	1981	
Magnus — FLAGS line Northern leg (spurs from Hutton, Heather, Murchison, etc.)	BNOC/ British Gas			or possibly part of gas gathering system	
Ninian — WELGAS line	Chevron	10	10	1980	
N.E. Frigg — Frigg	Elf	10.5	16	1982/3	
Odin — Frigg	Esso	14	20	1982/3	
UK gas gathering system: (quadrant 211 via Beryl via quadrants 15, 16 to St. Fergus)	Multi User	up to 572 miles, including spurs	main 36 spurs 16	1983/4	Outline planning permission received for BGC plant at St. Fergus
Brent — St. Fergus (FLAGS line)	Shell Expro	280	36	complete (commissioning Dec. '80)	Nearing completion
N. Cormorant — WELGAS line	Shell Expro	14	10	1980	
Cormorant — Brent (WELGAS)	Shell Expro	25	16	complete	
Frigg — St. Fergus	Total	225	32	complete	complete

Abbreviations: WELGAS = Western Leg Gas Line (Western leg of FLAGS)
FLAGS = Far North Liquid and Associated Gas System

SUMMARY   *COMPLETED*
32 inch pipeline — 225 miles   *Total completed = 530 miles*
36 inch pipeline — 280 miles   *Total planned to be completed = up to 1158.5 miles*
16 inch pipeline — 25 miles   *i.e. 31% of proposed system of pipelines offshore Shetland was completed by end 1980*

One option the UK Government can exercise if it wishes to limit the rate of production on these and other oilfields, and hence the rate of possible further developments, is to "bank royalty oil". This would happen if the State opted to take its 12½% royalty from the oilfields in oil rather than in cash, and decided to leave this oil in place to be collected at a later date. The practice of banking royalty oil could slow the rate of exploitation offshore Shetland, but will probably not to any significant extent, as this practice might prove inadvisable in some geological structures for technological reasons.

For further developments, the main value of the fields already in production or under development is that they will already have oil and gas pipeline systems* production and other facilities, which would allow the faster development of nearby smaller accumulations of oil and gas, through the tieing in of subsea completion wells. Such occurrences are likely, if not already happening, on Beryl, Cormorant, Heather, North West Hutton and Thistle Fields, (maybe eventually on all the Fields in Chart I).

(ii) Discoveries not yet with Department of Energy Approval (see Charts II and III) and new Discoveries

These discoveries could be more vulnerable to having their development plans postponed given certain policies and circumstances. It is perhaps useful to consider what circumstances would affect the rate of future development offshore, beyond those already mentioned which have or may have affected previous developments.** The main factors which might affect future developments would be:

(i)   The world wide political situation in countries holding oil and gas resources;

(ii) Government Policy;

(iii) The demand for oil;

(iv) The price of oil;

(v)   Technology — availability of men, machines and methodology and hence oil;

(vi) Finance.

(i) **The worldwide political situation in countries holding oil and gas resources** will affect all the other factors because the exploitation of these resources is conducted on a global basis; i.e. if one or more countries holding these resources in any quantity is torn by political strife and its normal supply of oil and gas restricted, there will be pressure put on the remaining countries holding resources to produce more. Conversely in a totally peaceful global situation, there might be pressure exerted on the producing nations to cut production in order to avoid a situation of excess supply thus reduce the price of the commodity.

(ii) A **Government policy** of conservation and possibly restriction of developments could seriously influence both immediate and the future production and exploration anticipated offshore Shetland. Many projects offshore are substantially financed by the oil companies themselves. If production is restricted, it slows cash generation, cash which might be used to undertake further exploration drilling.*** In the long term however, it would lengthen the 'Oil Era' but at the cost of a reduced rate of activity.

(iii) The **demand for oil** is to some extent reliant on Government pricing policy. It also depends on the state of the economy, i.e. in recession less oil is needed; the use of other fuels (see Table VI), and such unpredictable factors as mild weather. In 1979 the UK demand was 95 million tonnes.

See Tables Nos. IV and V on Oil and Gas Pipeline Status Offshore Shetland as at December 1980 which give an indication of the network of pipelines offshore Shetland which might be used for future developments once their current projected usage ends.

**Circumstances mentioned as possibly affecting some previous developments were:*

(a) a field with gas not having a pipeline system readily accessible when required;

(b) restrictions on flaring gas;

(c) not having a platform and other production facilities nearby;

(d) Government policy to 'bank royalty oil'.

****The cost of drilling a well is on average £5 million (1980 money).*

ENERGY PRODUCTION AND CONSUMPTION IN UK. 1979 and 1980 TRENDS. Table VI

UK ENERGY BALANCE (million tonnes of coal or coal equivalent)

CONSUMPTION (primary fuel input bases)

Year	Total	Coal	Petroleum	Natural Gas	Nuclear/Hydro
1979	355.9	129.6	139	71.3	16.0
1980	329.4	122	121	71.0	15.4
% change	− 7.4	− 5.8	− 12.9	− 0.5	− 3.7

PRODUCTION

Year	Total	Coal	Petroleum	Natural Gas	Nuclear/Hydro
1979	328.9	122.4	132.4	58.1	16.0
1980	337.4	130.1	136.8	55.1	15.4
% change	+ 2.6	+ 6.3	+ 3.4	− 5.1	− 3.7

Source of figures: Department of Energy

Comments: In 1980 total UK oil production was 80.47 million tonnes; consumption was 80.59 million tonnes, but in the last quarter of 1980 oil produced slightly exceeded oil consumed, so beginning the era of self sufficiency.

In 1980 this was estimated to have fallen to approximately 80 million tonnes.* It was not easy to predict future demand but demand from 1980 to 1990 has been estimated to be 80 to 95 million tonnes per annum (i.e. 1.6 to 1.9 million barrels per day). By the year 2000 the UK demand could range from as low as 70 million tonnes per annum, (1.4 million barrels per day) to as high as 100 million tonnes per annum (2 million barrels per day). There is a move throughout the world to reduce industrial fuel oil demands as far as possible and only to burn oil for transport needs, hence a possible move to alternative fuels is thought likely.

(iv) **the price of oil** if kept high by Government policy, will discourage consumption. At the same time the UK North Sea oil price is dictated by international factors, and in effect indexed to the price of, for example, Arabian or Libyan light crude.** The higher the oil price the more viable previously 'marginal' fields become to develop.

(v) **Technology.** It is suggested that there may be a shortage of the engineers, and various other skilled people needed to provide exploit and produce oil, within the next five to ten years. Currently machinery is stretched to the limit — for example, drilling rigs are in very high demand, and drilling tools and other equipment are similarly affected. At the same time new technologies are being demanded of the oil industry, for example, in order to develop the oil reserves west of Shetland (see Clair Field in Chart II). The rate at which these problems can be resolved will affect the rate of developments.

(vi) **Finance.** There is a limited world pool of money to finance oil and gas developments. If more attractive possibilities occur elsewhere, there could be a shortage of finance for some of the more risky North Sea developments.

---

*See Table VI "Energy Production and Consumption in UK".
**Most UK North Sea Oil is a light oil having high API rating of 30 degrees or more. See Glossary.

The price or cost of money is always an important factor in any development, and particularly in the North Sea where operating costs are high.

What then is to be the future, given some of these factors? It is estimated that probably all the discoveries noted here (Charts I, II, III) will be developed in the long term, and the future discoveries yet to happen. Shetland's Oil Era will continue for the forseeable future — which is to say, at least until the year 2000.

It has also been estimated,* that there will be a very great upsurge in developments within the next ten to fifteen years, doubling employment offshore, and hence affecting onshore facilities too. Whilst impossible to estimate the exact timings of these developments, it cannot be doubted that the situation for the future appears dynamic, for both the offshore developments and the onshore facilities supporting them.

It is thus to the onshore developments in Shetland and to how the Islands' people have managed these developments that this story now turns.

John M. Raisman "Oil and Gas — More to come from Scottish waters?" paper given to "Energy in the 90's Conference", Aviemore, 25th September, 1980.

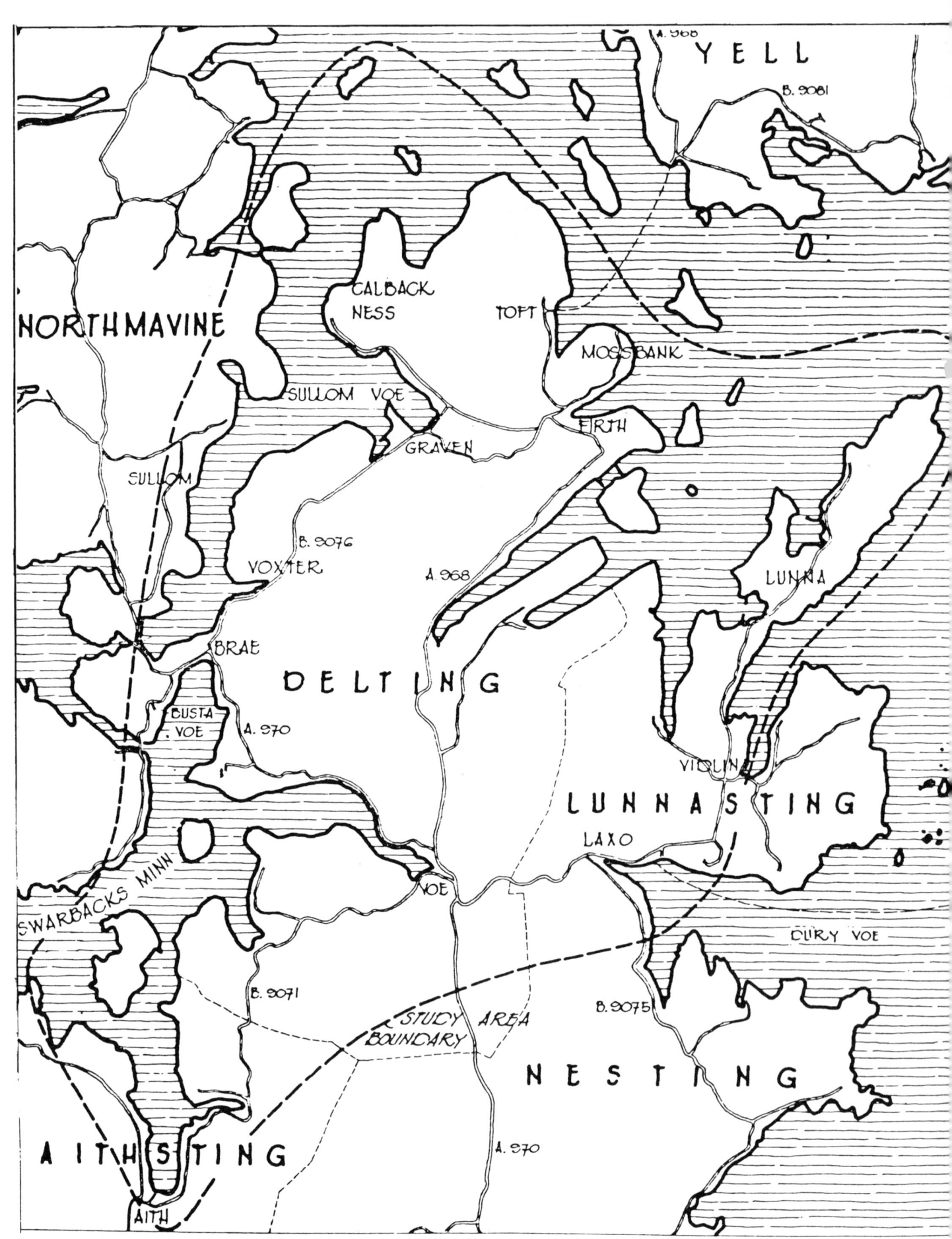

**ZETLAND COUNTY COUNCIL
INTERIM DEVELOPMENT PLAN STUDY AREA**

Chapter Two

Planning for oil

From the summer of 1971, the escalating pace of oil related activity in the waters around Shetland had aroused intense speculation, and some fears amongst the Islands' inhabitants. It became clear that these offshore activities would have onshore repercussions. The formation of a Council policy to control these developments onshore was reflected by a number of steps involving planning and legislation. Both planning and legislation were activities that were occurring almost simultaneously.

The Council began by assuming in early 1972 that major oil developments in the Shetland Islands were inevitable. To this end important moves in respect of future planning and legislation were made — viz:

A. Transport Research Limited were commissioned to make a study of deep water anchorages suitable for supertankers berthing at an Oil Terminal;

B. The Council instructed the County Development Officer to have an Interim Development Plan drawn up for the County. The purpose of this plan was to —

   i. establish a planning control framework to guide proposals for development throughout the County area.
   ii. identify priority areas for detailed survey work.

C. In April 1971, the Council decided to acquire, by Compulsory Order if necessary, land required by the Oil Industry. Before the Interim Development Plan was presented in draft form to the Council in October 1972, Shell/Esso made the announcement of the discovery of the Brent Field with potential recoverable reserves of 1000 million barrels of oil. This announcement confirmed Council convictions that major oil developments would occur. Shell indicated that apart from platforms on the Field offshore, a pipeline to carry the oil to tank storage facilities in Shetland would be required. These developments would result in a considerable influx of workers into Shetland, there being no surplus of indigenous labour. Against this backdrop, the draft Interim Development Plan was presented to the Council in October 1972.

Within a very short time of this plan being presented to the Council, problems had arisen.

A group of businessmen, totally independent of the Council, proceeded to register a company for the development of the Sullom Voe area as the 'Nordport Company Limited'.

The formation of the Nordport Company, and its actions of taking options to buy land around Sullom Voe, confirmed the Council in its determination to obtain control of these developments.

In November 1972, the Zetland County Council decided to apply for the promotion of a Provisional Order in Parliament, under the Private Legislation Procedure (Scotland) Act, of 1936. The Council required the authority to do the following:—

1. Acquire land for development.

2. Exercise harbour jurisdiction and powers in respect of areas liable to development — for example, Sullom Voe, Swarbacks Minn, Baltasound.

3. Licence construction works in these areas.

The Provisional Order was sought as a reinforcement of the Planning Acts. It would prevent speculators monopolising the land, thus inhibiting its best overall usage for the benefit of the Community. It would also allow the Council a continuing control over developments. Without additional legislation, such activities would have been beyond the powers of the Council as a local authority.

Local fears were expressed at this action of the Council, especially by those people whose lands were liable to be subject to Compulsory Purchase Orders. In December 1972, the Council issued a ten point policy document to alleviate such fears. This stated that the people in the areas concerned would be fully consulted before the Council took any action; good agricultural land was to be conserved wherever possible; where tenants had to be rehoused accommodation would be provided in areas of their choice.

In January 1973, the Council took further steps to provide itself with a full Planning Policy, by instructing the consulting engineers, Livesey and Henderson, to prepare a detailed survey of Sullom Voe. Meanwhile, the Nordport Company had applied for planning permission to develop the Sullom Voe area as an oil terminal. This company had either bought up or had options to 40,000 acres of land in and around Sullom Voe. Planning permission for such a venture was refused by the Council in February 1973, on the following grounds:—

1. The proposals were not in accordance with the policy of the Planning Authority;

2. The application was premature in that the Planning Authority had commissioned a study of the areas Sullom Voe and Swarbacks Minn, which study was to result in the production of a Master Plan and report for the areas, and would not be completed before 31st July, 1973;

3. The proposals were not sufficiently detailed. (Further details were asked of the Company by the Council on five different occasions. The only, somewhat sketchy, details provided were those in the application forms).

The Provisional Order had meantime been presented for Parliamentary consideration. The legislative process was furthered in March 1973, when the Lord Chairman of Committees, and the Chairman of the Ways and Means Committee in Parliament announced their decision that the Zetland County Council's objections should be dealt with and promoted by a Private Bill rather than by means of Provisional Order*. This the Council agreed to do.

In summary then, the Council in this short period of time had taken action:—

a. it commissioned Transport Research Limited to conduct a survey of deepwater anchorages;

b. it had prepared and submitted to the Secretary of State for Scotland an Interim Development Plan in March 1973 (the draft of this Plan went before the council in October 1972, as mentioned);

c. commissioned the consultants, Livesey and Henderson, to prepare a detailed Plan for the area surrounding Sullom Voe, which was completed in July 1973;

d. commissioned a detailed study by Geostock (a French registered company owned by Shell, BP, Total and Elf) into the feasibility of underground storage at Calback Ness (these costs were underwritten by the oil industry);

e. engaged the consultants Messrs Llewellyn-Davies, Forestier, Walker and Bor to prepare a County Structure Plan. This Plan was commissioned with the Scottish Development Department**;

f. engaged the consultants Llewellyn-Davies, Forestier, Weeks, Walker and Bor to prepare Local Plans for Unst and Yell; and the consultants Moira and Moira to do the Local Plan for Lerwick***, together with the surrounding areas;

g. established a Planning Department in September 1973.

The Lord Chairman of Committees and the Chairman of the Ways and Means Committee were of the opinion that the provisions of the draft Provisional Order related to matters outside Scotland to such an extent, and raised public policy matters of such a novelty and importance that they ought to be dealt with by a Private Bill.

**This Plan was finally handed over to the Council in 1975 and comprised a six volume report of Survey, and a two volume written statement — the draft Structure Plan itself. Due to the lapse of time between the initial survey, and the presentation of this material, the Council, as Local Planning Authority, decided that the Report of Survey should be updated and revised. This was completed in 1976, complying with the statutory requirements of Section 4 of the Town and Country Planning (Scotland) Act, 1972.*

***This original commission was for Gulberwick, Lerwick and Bressay. The Gulberwick and Bressay plans have been adopted but the Lerwick Local Plan is still in preparation but is expected to be adopted in 1981.*

Within this period, a further insight was gained into Zetland County Council thinking, when in reaction to a planning application from Shell for a £20 million super tanker terminal at Sullom Voe for crude oil, which would be transported by a pipeline from the Brent Field, the Convener, George W. Blance, stated — "The Policy Committee of the County Council are of the opinion that any such development should be by way of an oil industry/County Council partnership. Failing this, they consider that the facilities should be provided by the County Council. However, should the industry decide to proceed independently the Policy Committee are of the opinion that it must be on a 'joint-user' basis, and are of the opinion that the County Council should not process any such planning application until the Interim Development Plan has been accepted by the Secretary of State for Scotland, and the Zetland County Council Bill has been enacted. Furthermore they consider that the County Council require to be in a position to assess the difficulties which must be provided against in any planning permission before reaching decisions on any application".

Whilst the Private Bill began its somewhat stormy passage through Parliament, the County Council anticipated the Bill becoming law, and carried out the above noted policy statement by the following actions —

a. by seeking the advice of the National Ports Council on a suitable remit for its Ports and Harbours Committee, and on staffing (this was considered in December 1973);

b. by appointing a Ports and Harbours Committee;

c. by pursuing the question of pilotage powers with the Department of Trade and Industry;

d. by making contact with firms expert in providing towage services to discuss possible arrangements;

e. by notifying Shell that the County Council required immediate advice on the Company's requirements if facilities were to be available in 1976. (Such has been the change in time scale for development in Sullom Voe, from 1973 to the present time! Few people in 1973 fully appreciated the difficulties in developing the oil and gas fields in the deep waters surrounding Shetland).

The Council were correct in their belief that the Private Bill would eventually become law. In April 1974, the Royal Assent had transformed the Private Bill into the Zetland County Council Act 1974. The process had not been without delays. This is particularly so if one considers that it was in November 1972 that the Council decided to seek further powers to control oil related developments. Reasons for the delay were caused both by objections being raised to the Bill, and by unforeseen circumstances.

There was a somewhat unpromising beginning to the Committee stage of the Bill, which began in the Edinburgh Law Courts in June 1973, taking the form of a Public Enquiry before a House of Commons Select Committee. Both the Zetland County Council, as proposers of the Bill, and those who were objecting to the Bill, stated their case. The composition of the two sides was as follows —

1. Zetland County Council, represented by Mr G. W. Blance, the County Convener; and Mr Ian R. Clark, the County Clerk and General Manager.

    Appearing for the Council were:

    Livesey and Henderson, consultants;

    The Chairman of the Shetland Civic Society, Dr. Manson;

    The Chairman of the Crofters Union, Mr Laurence Graham (acting in his personal capacity);

    Mr R. H. W. Bruce, Lord Lieutenant for Shetland and Commissioner for the Crofters Commission;

    Mr George Hunter a member of the Countryside Commission for Scotland;

    Mr W. George for BP; and Mr John M. Drummond for Shell.

2. **Objectors to the Bill:**

    Residents of Graven and Baltasound;

    The Nordport Company Limited;

    Total Oil Marine;

    Lerwick Harbour Trust;

    Blacksness Pier Trust;

    The Shetland Fishermen's Association.

The hearings in Edinburgh began, but were interrupted whilst the four members of Parliament on the Select Committee visited Shetland on 14th June, and returned to Edinburgh on 16th June after taking evidence from supporters of the Bill. The sitting of the Committee recommenced in Edinburgh on 18th June.

Many of the objections were erased by minor alterations to the Bill. Strongest of the objectors was the Nordport Company — perhaps as the one that had most to lose. If the Bill went forward in any form, which included the Compulsory Purchase Powers that it gave the Council, it would have meant Nordport losing all chance of supplementing its prospects to develop the land around Sullom Voe for which options had already been taken*. The Select Committee, however, ruled against the Zetland County Council, on the question of Compulsory Purchase Orders.

The County Council refused to accept this as the final decision on the subject. By judicious lobbying, and explanations of their intentions to Members of Parliament, the Council finally won the day. Several factors helped them in this instance including:—

1. The Arab threat to cut off the Middle East Oil made the speedy development of North Sea Oil imperative;

2. Livesey and Henderson had presented their report in July; this allowed the areas of Compulsory Purchase Orders (due to oil-related developments) to be reduced dramatically. Land around the Baltasound area was omitted from the projected plans entirely, and that around Graven reduced from one area of 9700 acres to three small areas of 2769 acres in total.

3. A change of Government in 1974.

The Bill was then presented to the House of Commons for consideration in this amended form, in November 1973.

In the midst of these events, on the 18th December, 1973, a special meeting was called of the Zetland County Council to meet Lord Polwarth, the Minister of State at the Scottish Office, and representatives from the Scottish Economic Planning Department, and Scottish Development Department.

The Convener of the Council, Mr G. W. Blance, recalled the steps taken by the Zetland County Council to deal with oil developments in Shetland. He stated that the County Council had a responsibility to protect Shetland and the people of Shetland, and referred to the fact that earlier in the year decisions had been made —

1. that major oil developments should be concentrated at Sullom Voe;

2. that there should be an integrated development there;

3. that the oil companies be asked to consider that the facilities should be organised on a joint user basis.

Mr Blance went on to note that the three main parties responsible for the fulfilment of the Council's policy were —

    the County Council,
    the Oil Companies, and
    the Government.

The Convener stated that the oil companies must realise that they must come into Shetland and proceed in conformity with the County Council's policy, and that they should come as an industry, and not as individual bodies; and that the Government must realise Shetland is a small county and that there was a great risk of violent disruption to the present way of life in Shetland unless everything was planned; and that the Government must "bend" regulations if necessary to assist the Council. Mr Blance also stressed that the Government must make financial and technical assistance available to enable the Council to provide services such as roads, water, drainage, educational and recreational facilities. The Convener also referred to the Council's present labour problems, and stated that they would welcome advice and assistance to overcome the difficulties arising out of differentials in wages.

In reply Lord Polwarth then made reference to the additional oil strikes which had been announced in the East Shetland Basin in the previous few weeks, and stated Shetland was in an important position and it was clear that oil would be piped to Shetland where it would be stored, subjected to certain processes and re-shipped. He said

---

It has subsequently been demonstrated that the main beneficiaries from land sales at Sullom Voe were the speculators and not the farmers/crofters. For example, land which the speculators acquired at £200 per acre from local farmers, was acquired by the Council for a considerably greater sum.

that it was important to plan flexibly for development and gave assurance of help from the Government and the Scottish Office to enable the County Council to play a central part in this operation.

A discussion ensued where the Convener referred to the past projects for schools and other works that had been submitted to Central Government Departments with the answers being that there was no money available to back the projects. Mr Blance enquired whether the Council would again get such an answer when schemes were put forward in the future. Lord Polwarth indicated that if any project could be shown to be connected with North Sea Oil, then priority would be given.

Meanwhile, whilst these matters of planning policy were being agreed between the Government and the Council, the Private Members Bill continued its progress.

By the time it has passed its Second Reading in the House of Lords, on the 17th January 1974, only the Nordport Company remained as objectors.

An unforeseen factor caused the Bill to be delayed from progressing further — the calling of a General Election, which caused Parliament to be dissolved. Fortunately for Shetland's subsequent history, on the same day as the Parliament was dissolved a carry-over motion was passed, which allowed the Bill to be reintroduced in the new Parliamentary session on 12th February 1974*.

The Third Reading of the Bill was in the House of Lords on 26th March 1974, the House of Commons considered the Lords' amendments on 2nd April 1974, and the Bill became law as the Zetland County Council Act, when given the Royal Assent on 10th April 1974.

As the history of events implies, it was the threat of a sudden unknown, and potentially disruptive series of oil developments which served as a spur in driving the Council to have the Act passed.

Nevertheless, it must be noted, that whilst the Council gained powers which would enable it to have greater control over oil related developments, the Act was drafted in terms of a wider vision. It was designed to give the Council powers which would encompass Shetland as a whole, and which would out-last the Oil Era — for example, powers of coastal conservancy.

It would be fallacious to consider this statute purely in terms of the powers it gave the Zetland County Council; in keeping with the spirit in which all legislation is enacted, Parliament, whilst granting certain powers to the beneficiaries (in this case the Zetland County Council) imposed certain duties and obligations upon them.

Powers given to and Duties imposed upon the Council by the Zetland County Council Act 1974 (with particular respect to oil related developments)

Within the context of this publication, it is only the Zetland County Council's powers and duties as applicable to oil related developments which are considered relevant. These could perhaps be summarised in the following manner —

1. **Harbour Authority** — The Council can exercise jurisdiction as a harbour authority** and powers as Harbourmaster can be exercised in the Sullom Voe and Baltasound areas. Ports and harbours excluded from the Act were those within the jurisdiction of Lerwick Harbour Trust, Broonies Taing Pier Trust at Sandwick, and Blacksness Pier Trust at Scalloway***. For example, being a Harbour Authority allows the Council to control the marine activity associated with the oil terminal at Sullom Voe.

It is only certain Private Bills that are eligible for carry-over motions of this kind; had this carry-over motion not been passed, the Bill would have died with the old Parliament, and would have had to be reintroduced with considerable consequent delays.

**The powers of Harbour Authority as confirmend by Act of Parliament were not unique to the Zetland County Council Act. For example, The Zetland County Council Symbister Harbour Order Confirmation Act, 1961, gave the Council powers as harbour authority for Symbister; it also gave powers to the Council to acquire land in the vicinity of Symbister Harbour.*

***Blacksness Pier came into Shetland Islands Council ownership in 1976, and is now administered by a joint Management Committee of Shetland Islands Councillors, and of local Scalloway and Burra Community (mainly pier users) representatives.*

2. **Construction of Works and Dredging** — The Council may issue licences to dredge and licences to construct works within the three mile limit of the coastal area, outside areas within the jurisdiction of Lerwick Harbour Trust, Broonies Taing Pier Trust and Blacksness Pier Trust. For example, the Council has subsequently exercised control by considering applications for construction and other jetties at Sullom Voe, pipeline works*, moorings for oil related structures, fish farming equipment, and permission to dredge sand and aggregates.

3. **Powers to invest in Bodies Corporate** — The powers given to the council by the 1974 Act enabled it to invest in 'bodies corporate' or companies such as the Sullom Voe Association Limited, Grandmet (Shetland) Limited, and other business ventures. (These are mentioned in more detail in Chapter five of this book).

4. **Control of Development** — The Council has the control of the conservancy (eg. environmental conservation) and control of development in the coastal area (within three mile limit) and the vicinity of a harbour area.

5. **Compulsory Purchase of Land** — The Council has the power to acquire land for oil related development within the designated area (the three areas of 2769 acres around Graven). This power to give the Council ownership of the development land is the corner-stone to the control of oil developments. However, before land can be compulsorily purchased, the approval of Parliament has to be sought and gained in the form of a statutory instrument. This makes Shetland somewhat unusual, as in Scotland local authorities normally only have to gain the consent of the Secretary of State to compulsorily purchase land. The Council have, in fact, had two Compulsory Purchase Orders (CPO's) confirmed under the Zetland County Council Act procedure.**

6. **Shipping** — The Council may construct, purchase, or contract for hire, vessels required by it for carrying out its function under this Act. These functions include the towage of ships. For example, this section of the Act was used to allow the creation of Shetland Towage Limited, the Company established by the Council in conjunction with towage companies, to own and operate tugs at Sullom Voe.

7. **Reserve Fund** — The Council may create a Reserve Fund which may be used for various harbour purposes or for any purpose which in the opinion of the Council is solely in the interest of the Shetland Islands or their inhabitants.

However, the fund can only be created when the Harbour account shows a profit. At the time of writing this had not happened, but is expected to in the near future.

The above were some of the powers which the Zetland County Council (and its successor, the Shetland Islands Council), gained under the 1974 Act. One of the objectives of this Act, as stated, was to have a continuing control over oil related developments. It is to these developments that this publication now turns, noting where relevant, the implementation of the Councils planning policy and the use of the available legislation to ensure that the best interests of the community and its special environment are protected.

Both the Ninian and Brent pipelines had to be licensed by the Shetland Islands Council within the three mile limit, in order for the construction to be carried out.

**One was for the major part of Calback Ness, the other for a road to the Port Administration buildings at Sella Ness. The major part of Calback Ness was eventually acquired by statutory conveyance and the Sella Ness area is also being acquired in the same manner, thus the CPO's have never required to be enforced.*

Chapter Three

Marine Supply Services
Supply Boats and Supply Base Developments

BACKGROUND

One of the first signs to be seen in Shetland of the impending oil and gas developments offshore was the appearance in the Islands' harbours of supply boats and other oil related craft needed to service offshore vehicles involved in exploration activities. By 1972, it had become clear that with major discoveries occurring offshore, Shetland would need supply bases. The logistical argument for this is that the closer a supply base is to the field, the fewer supply boats needed per period of time, to make the required trips.* Thus in May 1972, planning permission had been sought and obtained from the Zetland County Council for the construction of the first of the supply bases at Broonies Taing, Sandwick, (15 miles south of Lerwick).

The Ninian Field and other major discoveries had occurred after the Brent Field (see Chapter I), all within a comparatively short space of time. After the establishment of these discoveries there has followed the spectacular and simultaneous development of several major fields, with the drilling, construction and production activities going on continuously.

The Activities Serviced by the Supply Boats and the Type of Cargoes Carried

The supply boats have been servicing the vehicles which are required for offshore activities, namely:—

1. Mobile drilling rigs.
2. Fixed platforms.
3. Pipelaying barges.
4. Bury barges, to bury the pipe which has been laid.
5. Derrick barges, and vessels conducting offshore construction, hook up of systems, and maintenance operations.

This situation has put an enormous pressure on the supply of materials which had to be achieved at nothing less than 100% instant response even at a premium cost. As a result, four more supply bases have been established in Shetland, since the first base at Sandwick was completed in 1973. These bases are most likely to service supply boats operating north of latitude 59 degrees North in the UK North Sea. Fields south of 59 degrees North, for example, tend to be serviced out of Aberdeen. Fields straddling the UK/Norwegian boundary line, where the majority of the resource is in the Norwegian sector, for example, Statfjord, tend to be serviced from Norway. Supply boat activity and supply base developments have tended to reflect directly the three major long term** activities offshore, which can be categorised as:—

See Table VII "Comparison Between Supply Boats Operating from Lerwick as opposed to Aberdeen to Fields Offshore Shetland". A further comparison was made in a study done for Unionoil, as operators of the Heather Field. This showed optimum steaming time from Aberdeen to Heather was 27 hours and from Lerwick to Heather seven hours. Sailing from Lerwick made a net saving in terms of fuel costs of £4860.60 over sailing from Aberdeen, not taking into account the cost of chartering a supply boat at £3000 per day. See "An Investigation into the Feasibility of Setting up an Oil Supply Base in Shetland" by T. J. Little of Robert Gordons Institute of Technology, Aberdeen, 1980.

**These activities are considered long term in the sense that they are likely to be continuous in the area offshore Shetland until the end of this century or beyond, even if they only last a short time on each individual field. Pipelaying is not included here as it is a sporadic activity.*

Table VII
A COMPARISON BETWEEN A SUPPLY BOAT OPERATING FROM LERWICK AS OPPOSED TO ABERDEEN — TO FIELDS OFFSHORE SHETLAND

Supply Boat "Mercia Shore" — Cargo Ex. Lerwick — 82 Sailings

	Metric Tonnes
Fuel	5819
Water	7177
General	5735
Tubulars	3615
Bulk Chemicals (Barytes, etc)	7243
Brine/Mud	7634
Tonnage Moved	38223
Average cargo per sailing:	466.13 tonnes

Supply Boat "Edda Sprite" — Cargo Ex. Aberdeen — 44 Sailings

	Metric Tonnes
Fuel	6492
Water	4053
General	4193
Tubulars	2334
Bulk Chemicals (Barytes, etc)	1192
Brine/Mud	292
Tonnage Moved	18556
Average cargo per sailing:	421.72 tonnes

Source of information: Shell UK Exploration & Production Limited.

Graph D: Distances between Principal Supply Bases and Oilfields off Shetland

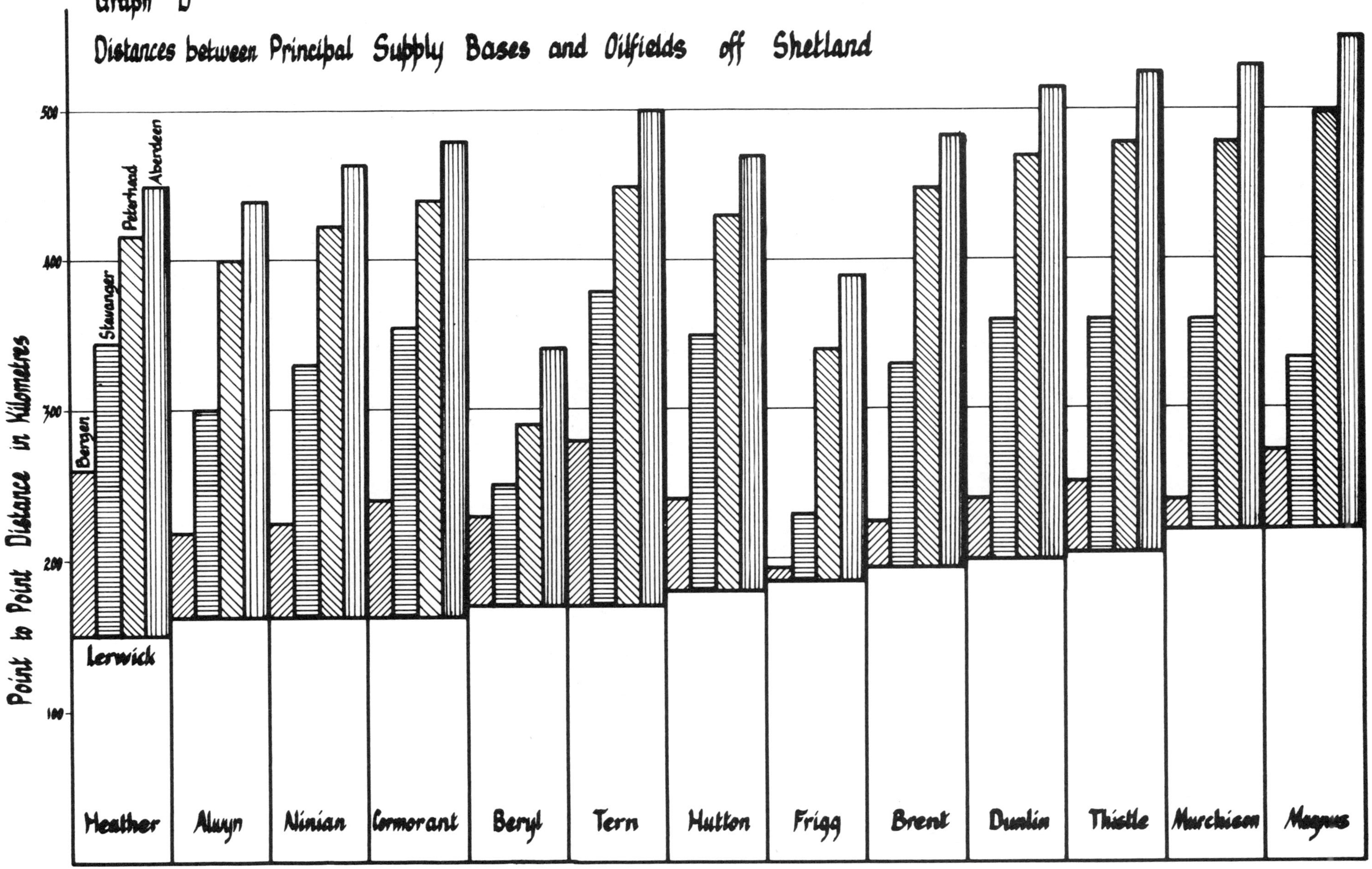

(Source of Information: Norscot Oil Services Limited)

A. Exploration/appraisal Drilling

B. Construction/Hook-up of platforms

C. Drilling, Production, Maintenance and Engineering on a platform after installation

A. Exploration and Appraisal Drilling

This activity is usually conducted in the northern North Sea by semi submersible mobile drilling rigs*, which are either self propelled and manoeuvrable by their own power or are partially powered units requiring the assistance of anchor handling ships and/or tugs. Once the rig is moored, the anchor handler or tug is no longer required as long as weather conditions remain within operational limits. If the operators' anchor handling/rig moving requirements during a prolonged period (e.g. one year) are going to be continuous, then these vessels need to be able to perform a supply function between anchor handling and rig moving activities, for a long term charter to be economical.

This exploration/appraisal drilling stage requires the full range of materials except specialised engineering items. Materials required will include: tubulars, wellheads, handling and drilling tools, chemicals, cement, general and life support materials. The type of vessel required for this work will be versatile, being able to perform the duties of platform supply/pipe carrier, anchor handling and towage**.

Deck cargo consists of: Tubulars, drilling and handling tools; wellhead equipment, chemicals and food (containerised), aviation fuel (600 gallon tanks), general equipment.

Under deck cargo consists of: Fuel, water (drinking and drilling), liquid mud, bulk cement, brine, chemicals. Quantities of cargo required are relatively moderate under the normal progress of drilling, but well conditions may demand high offtakes in both bulk and palletised chemicals at irregular intervals during exploration and appraisal. Each voyage will be up with a different cargo depending on the state of the well. However, the following gives an average monthly breakdown of tonnage shipped to a single rig in exploration:— Fresh water — 970 tonnes; gas oil — 460 tonnes; bulk chemicals — 360 tonnes; tubulars — 210 tonnes; general cargo — 200 tonnes; cement — 145 tonnes; bagged chemicals — 100 tonnes. This gives a total of 2445 tonnes per month.*** In general, it could be said the needs of a rig in the exploration and appraisal stages are greater than those of a platform at the production stage.

In general, it could be said the needs of a rig in the exploration and appraisal stages are greater than those of a platform at the production stage.

A platform at the production stage has only 60 percent of the cargo requirements of a drilling rig engaged in exploration. A rig engaged in appraisal drilling may have slightly less requirements than one wildcatting**** on unknown structures. To generalise is perhaps unwise, but it is useful to have and "average" or "normal" figure to refer to.

B. Construction/Hook-up of Platforms

The construction and hook-up of a platform normally lasts approximately one year, with the emphasis on supplying engineering materials to complete life support (accommodation, power utilities, safety and firefighting facilities) and drilling needs. A peak workforce is usually accommodated initially on a gangway-connected dedicated floating hotel (normally a semi submersible vessel) — and later on the platform itself. Supply requirements are also high for food, general materials and water.

Thereafter, during platform drilling and subsequent production a considerable amount of engineering work is still carried out for a period of one year or longer, in order to complete gas/oil separation, oil pumping, gas and water injection, etc.

Manning levels are above the so called steady state level (the level compatible with accommodation facilities on the platform, usually $\pm$ 160) and need the deployment of the 'flotel' or accommodation unit. In the construction and hook-up of platforms anchor handling is not required unless supporting accommodation units have to be relocated. Underdeck capacity for supply boats is not essential but deck space, preferably protected against damage by seawater, is a prime requirement, given that much of the material, although crated, is of a mechanically delicate nature. Prefabricated materials, small accommodation units, engineering construction and scaffolding materials have to be transported. The platform supply/pipe carrier type is considered the most

---

Drillships are occasionally used, but usually for special type of work e.g. in 1980 the drillship 'Petrel' worked in 2,000 ft. water depth west of Shetland.
**See diagram for illustration of supply boat; also table VIII.*
***Source of figures: BP Petroleum Development Limited.*
****See glossary.*

suitable vessel for this function. Two supply vessels would probably be used in this situation — one dedicated to transporting the construction project materials, the second being on scheduled runs with food and other life support materials.

As the construction progresses, the average cargoes delivered by a project dedicated vessel decrease gradually. It then becomes feasible to incorporate project materials in a normal supply boat schedule and reduce the number of sailings needed accordingly.

A typical cargo carried by a supply boat to a production platform would be:—

Item	Tonnage
Gas Oil	1000
Deck Cargo	250 (mainly containers)
Fresh Water	1750
Bulk Chemicals	100
Total	1525 tonnes/voyage

Source of figures: BP Petroleum Development Limited. Derived from a study of "Stena Piper" on regular runs to Forties Field. A further point to note is that whilst this may be the cargo carried, it may not be the cargo unloaded, due to weather conditions. Also there may be more than one platform to be supplied by the same ship.

Other regular supply activities include the servicing of specialist craft used for underwater inspection and maintenance. Their supply needs will include fuel, water, food, diving gas and general diving equipment. These specialist craft are usually located in areas where platforms are operating and are included in the supply boat's 'milk run' or regular servicing schedule. To compare how the offshore activities of exploration and appraisal, construction and hook-up, drilling, production and engineering, have different supply needs one can perhaps compare cargo delivered to a drilling rig, with that delivered to a platform in production in 1979:—

Cargo Items	Drilling Rig (Tonnes) "Stadrill"	Production Platform (Tonnes) "Brent Bravo"
Fuel	3757	9070
Water	1647	1662
General	4465	5584
Tubulars	3965	4807
Bulk Chemicals	7235	3548
Brine/Mud	4777	906
Total (Metric Tonnes)	25846	25577

Source of figures: Shell UK Exploration and Production.

Platform Drilling/Production, and Drilling and Ongoing Engineering

A more balanced utilisation rate of supply vessel carrying capacities, underdeck, as well as ondeck, can be achieved once the production phase has been entered. During the production phase, activities one could expect to be occurring simultaneously would be:— drilling, well completions and workovers*, production and ongoing engineering covering new or replacement equipment, scheduled maintenance, and some other improvements from new equipment being commissioned. In respect of supply boat requirements, the emphasis will remain on deck space capacity for drilling, production, engineering and food supplies. Tubulars, drillpipe, test strings, and long length tools and piping, will utilise approximately 30 percent of available deck tonnage. Liquid oil based drilling mud can only be carried in supply vessels with converted tanks and pump systems, whereas light brine can be handled by most vessels and heavy brine by vessels with special tanks.

The major differences to be noted here are in the requirements for fuel, bulk chemicals, brine and mud. Such differences in emphasis allow supply boats of different design to be utilised — for example a supply boat for a drilling rig may make more use of underdeck storage tanks than one for a production platform where above deck storage space is at a premium.

Changes in the use of Supply Boats

Having described the activities the supply boats support, it is perhaps appropriate to describe briefly how the operations are conducted and the ships managed, because there have been considerable changes occurring in the decade or so since operations commenced offshore Shetland, which have affected the pattern of activity.

The supply boat industry consists of companies who charter their ships and crews to the major oil companies.

Charter arrangements split into long term (up to two years) charters for ships, and the short term or spot charter market when a ship may be chartered for one voyage or more. The kernel of the business is the long term charter market — this suits both the oil industry in that they can obtain better rates and of course the shipowners. Changes have occurred in that more boats are on long term charter than previously now that the production phase has arrived.

A charter for a supply boat was costing some £2,100 per day at July, 1980, with the cost of fuel adding about £900 per day.

General criteria used in deciding on supply boat scheduling and chartering are:—

(a) **The vessel's deck area**

   Not all the deck area is always useable; there can be limitations on the shape of a space and its accessibility.

(b) **The background of the ship itself and its owner/operator.**

   It is important that the ship is efficiently operated although it must be noted that the chartering company exercises a great deal of control over the operation of the ship, even to the extent of dictating the speed at which the ship will sail (under normal weather conditions).

The Number of Boats Used

In 1971, when supply boats first appeared in Shetland waters, it was common for two supply boats to be assigned to each rig, in addition to standby boats and anchor handling boats. Similarly, when the first fixed platforms were installed by 1975/1976, two supply boats were dedicated to each platform. By 1980, this practice had been radically altered, with the ratio being progressively reduced. For example, vessel design had improved, to allow vessels to combine the activities of supply/anchor handling and towage (in some instances). Further, where an operating company has more than one installation in an area (e.g. Ninian or Brent Fields), economies have been made in that the same supply boats may discharge at several different installations. Changes in design have also increased the size of the supply boats from those of ten years ago, in their above deck space and below deck (usually tanks) carrying capacity. The difference in the numbers of supply boats needed per installation could be noted as follows:—

See Glossary.

Vehicles Supplied	No. of Supply Boats Needed	
	1974	1980
Drilling Rig	2	1.3
Platform	2	1

Source of information: BP Petroleum Development Limited; Shell UK Exploration and Production Limited.

Scheduling and Loading Practices

Improved scheduling (schedules are usually calculated at least one week in advance) and loading practices have also helped reduce the number of boats required per installation and the number of sailings needed to supply each installation.

(a) **The scheduling system** is based on making frequent (up to three times per week) visits to each location, depending on the distance from port. This is useful also from the point of view of offshore loading, which is frequently subject to weather constraints (usually wind will cause a platform's crane to stop operation, in order to prevent shock loading, well before the highly manoeuvrable supply boat has to stand off). The scheduling system also tries to put together locations needing servicing in a geographical block, so that they can be serviced from one sailing.

(b) **Loading practices** have also been considerably altered since 1974 when a changeover to containerisation of deck cargo occurred. This cut losses of deck cargo dramatically, for as much as 90 percent of a deck cargo (drums, pallets) had been known to be washed overboard in a winter's sailing. Now, specially designed containers with internal bracings and tie-down points mean that cargoes arrive intact. A new industry in producing and servicing containers has developed — containers which can cost several thousand pounds each, especially if they are designed for delicate cargoes (e.g. computers).

SAILING DIRECTIONS

The greatly increased cost of fuel (bunkers) has led some oil companies to instruct vessels not to exceed 60 percent full power (approximately 11 knots for most boats) unless ordered. This has had a marked effect of the total fuel consumed by supply boats. For example, in 1979 when BP Petroleum Development Limited first instituted this order the figures were as follows:—

Months	April	May	June*	July	Aug.	Sept.	Oct.
Tonnes consumed	1364	1378	724	796	843	886	872

The master is obliged to provide the charterer with a form stating the time and distances steamed, and the fuel he has consumed. The major supply boat charterers such as BP and Shell have shipping co-ordinators to co-ordinate all their North Sea operations and practice forward planning for multi-destination voyages as far as possible.** This has resulted in both a reduction in fuel consumed and in the number of ships used per operation. It has also meant the supply boats used spend more time at sea, and less in port.

All the above practices and changes in practice occurring in the operations of supply boats have had their consequences onshore in Shetland and having thus briefly explained the position offshore, it is now the onshore position that must be noted.

---

*60% full power instructions instituted, combined with some reduction also in size of fleet. (Source of information: BP Petroleum Development Limited).

**Multi-destination voyages are invariably planned, except where just one rig is being supported away from the main routes.

Diagram I

Typical pipecarrier, platform or supply vessel for the North Sea

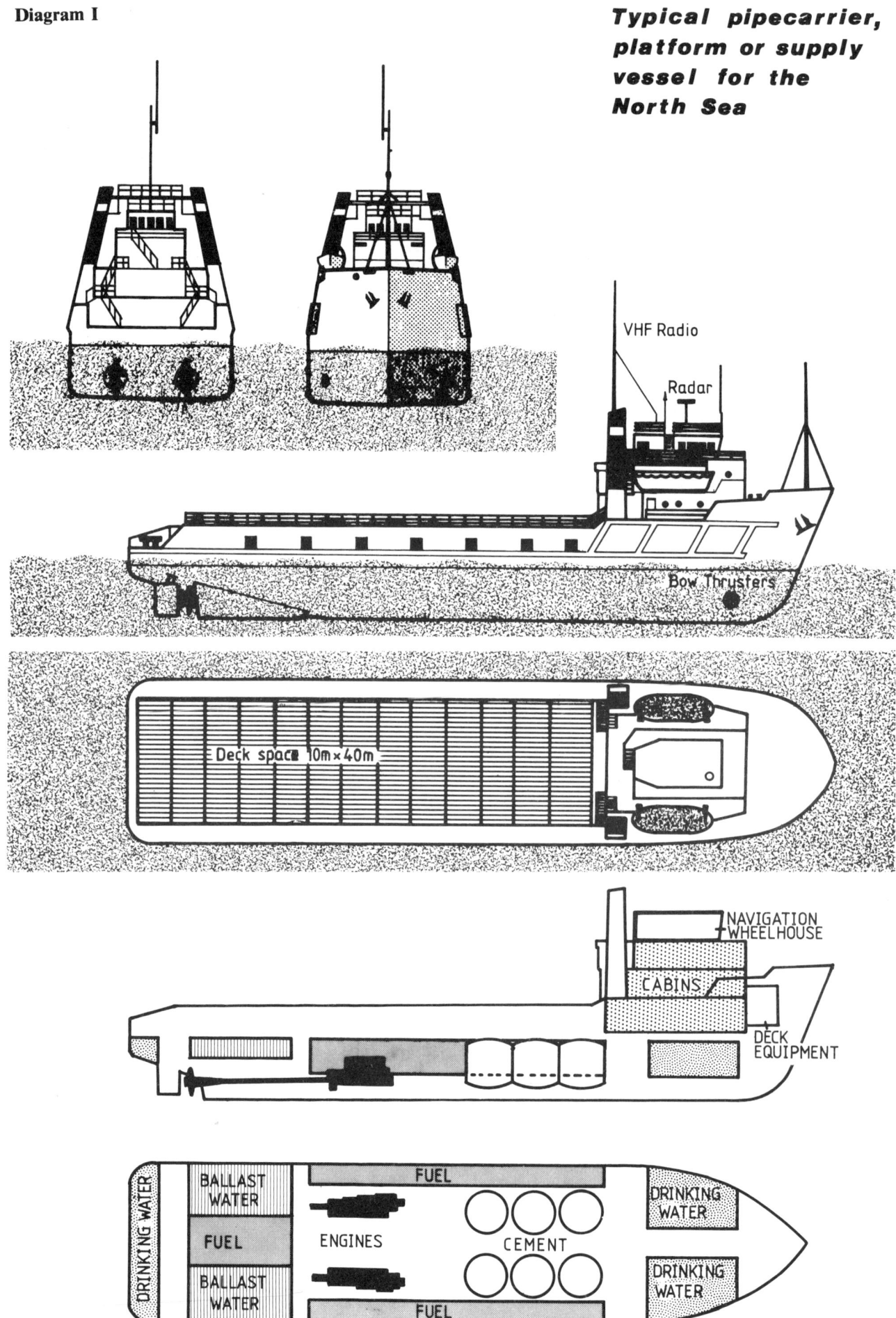

DEFINITIONS Table VIII

1. **Supply Boats**

   Specially designed vessels, with carrying capacity from 1400 (older designs) to 2500 tonnes or more, clear deck area being at least 45 x 15 metres, if not larger.

   Functions

   (a) Carrying food, technical supply requirements.
   (b) Towage operations — e.g. assistance in towing semi-submersible rigs to new positions.
   (c) Anchor handling — carrying and laying 15 tonne anchors to keep rigs in position; also for lay barge movements. Anchor handling is also done by specialised vessels.

   Crewing

   Two crews of (i) at least ten men for servicing production platforms.

   (ii) 12 men for anchor handling.

   Crews work rota of six weeks on duty, six weeks off duty.

2. **Research Vessels**

   Of special design for (a) underwater inspection and diving activities, perhaps using diving bells or miniature submarines, and utilising moonpools*; (b) seismic surveying.

3. **Standby or Safety Vessel**

   Originally of trawler type, over 30 metres in length, capable of accommodating on a short-term basis, the maximum number of people stationed on a platform at any one time. Now specially designed emergency support vessels have been built (see text) to withstand more severe conditions.

For explanation of term moonpool see Glossary.

OIL SERVICE VESSEL VISITS TO SHETLAND[*1]

Chart V

Year	Lerwick[*2]	Sandwick	Scalloway	Total	Comments
1971	79	—	—	79	This year marked the beginning of oil related activity in Shetland waters.
1972	151	—	32	183	
1973	328	34	1	363	
1974	838	110	16	964	
1975[*3]	1581	99	7	1687	Lerwick Harbour figures: 1075 supply boat visits; 373 research boat visits; 133 standby boat visits.
1976	1679	114	16	1809	Lerwick Harbour figures: 1305 supply boat visits; 268 research boat visits; 106 standby boat visits. Actual no. of supply boat vessels 123. Frequency of visits from one to 79 times.
1977	1724	53	40	1817	Lerwick harbour figures: 1409 supply boat visits; 195 research boat visits; 120 standby boat visits.
1978	1827	27	25	1879	Lerwick Harbour figures: 1529 supply boat visits; 211 research boat visits; 87 standby boat visits. Oil service vessels accounted for 43.16% of all vessel visits to the harbour.
1979[*4]	1531	N/A	7	1538	Lerwick Harbour figures: 1291 supply boat visits; 175 research boat visits; 65 standby boat visits. Oil service vessels accounted for 41.56% of all vessel visits to the harbour.
1980	1586	6	20	1612	Lerwick Harbour figures: 1383 supply boat visits; 153 research boat visits; 50 standby boat visits. Oil service vessels 42.27% of all vessel visits to the harbour.

NOTES:
  [*1] Not all oil service vessel visits are to pick up or deliver cargo; some boats visit to shelter from bad weather, some for repairs.

  [*2] Lerwick Harbour figures comprise supply, research and standby boats.

  [*3] Start of pipelaying operations offshore Shetland was partial cause of the 1974/75 increase.

  [*4] Decrease in traffic volume in 1979 was due to a fall-off in exploration drilling; possible also increase in the price of fuel revised scheduling practices and reduced the number of supply boats used to service operations.

SUPPLY BOAT BERTHS IN OPERATION AND WITH PLANNING PERMISSION   Chart VI
(as at January 1981)

Operator	Location	Berths In Operation	Berths With Planning Permission	Remarks
Hudsons Offshore Holdings (BP Minerals)	Sandwick	2		In early 1980 BP Minerals bought Selection Trust who owned Hudsons Offshore
BP	Lerwick	2*1		
Nordic Investments	Lerwick		2	Subject to Section 50 agreement. No N.P.C. Licence*2
Norscot	Lerwick	9		
Ocean Inchcape	Lerwick	3		
Shell	Lerwick	2		
Northern Offshore Services Ltd.	Scalloway	1		In use intermittently. Not a specialised berth*3
Offshore Oil Services	Basta Voe, Yell		2	
Nordic Investments	Baltasound, Unst		1 (1st phase)	Subject to Section 50 agreement. No N.P.C. Licence*4

Total: 19 berths operational, five with planning permission but as yet undeveloped.

NOTES:

*1 *BP at Lerwick have an agreement with Lerwick Harbour Trust to use an extra berth at Holmsgarth for supply boats, although this is not a specialised supply boat berth, and does not, as such, require planning permission.*

*2 *The N.P.C. or National Ports Council exercise jurisdiction over the granting of licences to jetties/quays, and normally only do so once a firm contract is awarded to the prospective developer.*

*3 *This berth is part of harbour facilities at Blacksness, Scalloway, owned by Shetland Islands Council.*

*4 *Nordic Investments are required to develop in Baltasound first, before being allowed to proceed with their proposal for Lerwick.*

THE PLANNING OF SUPPLY BOAT BERTHS IN SHETLAND

As stated, considerable changes have taken place since the early 1970's when the first oil service vessels visited Shetland and supply boat berths were planned to accommodate them. Thus, in 1978, there was existing planning permission for up to 32 berths, although only 19 were in operation.* Since then there has been a decline of approximately 15 per cent in the number of oil service vessel visits to Shetland.** This decline has been partly due to a decline in exploration, but also as noted to such things as the rising price of fuel, the resulting slow steaming, and improved scheduling and usage of ships by the oil companies — perhaps also the increased size of the modern supply boat has had some effect.

Since 1978, the planning permission for eight of the 13 undeveloped berths has lapsed, and the supply base at Sandwick has reduced its functions, although the two berths remain there.

It is anticipated that general use berths at Lerwick, Baltasound and Scalloway harbours will be used by oil service vessels increasingly as developments continue to progress. It is also possible that one or two more specialised supply berths might be built as both exploration and development traffic builds up again and projected developments materialise (see Chapter One).

THE SUPPLY BOAT BASES

There are presently five supply boats in operation in Shetland, namely: Hudsons, at Sandwick, and four others — BP Petroleum Development Limited, Norscot Oil Services Limited, Ocean Inchcape (Shetland) Limited and Shell UK Exploration and Production, all concentrated at Lerwick Harbour. Two fundamental categories of supply base exist:—

Category One

A base using existing port facilities ie. docks, warehousing, workshops, etc. In this case, only minor changes in existing facilities or installations are necessary, for example, the erection of silos for drilling mud and cement, and possibly the installation of facilities to provide bunkering fuel and fresh water. e.g. Ocean Inchcape previously operated under Category One until a purpose built supply base was constructed.

Category Two

A purpose built supply base. For such a base, an area may require to be levelled to provide new quays, warehouses, office buildings and suitable workshops. e.g. Norscot Base.

These categories of base are of the same nature in the services they provide to shipping. Both categories of base have existed in Shetland, although at present, only purpose built bases exist.

A further distinction can be drawn between supply boat bases in terms of their users —

1. **Multi User Bases**

    Bases administered by a company to provide services to all comers on a strictly commercial basis, can be denoted as Multi User Bases.

2. **Single User Bases**

    Bases administered by a company for its own exclusive use can be denoted as Single User Bases.

---

*See Chart VI "Supply Berths in Operation and with Planning Permission (as at January 1981)".
**See Chart V "Oil Service Vessel Visits to Shetland".

Bearing in mind both the Base Category, and type of user, the five Shetland Bases can be tabulated as follows:—

	Base	Type	User
1.	Hudsons	Category 2	Multi User
2.	Norscot Oil Services Ltd., Lerwick	Category 2	Multi User
3.	Ocean Inchcape (Shetland) Limited, Lerwick	Category 2	Multi User. Began as Category 1, but felt that business justified the provision of a purpose built base, commissioned in August, 1978.
4.	BP Petroleum Development Ltd., Lerwick	Category 2	Single User
5.	Shell UK Exploration & Production Ltd., Lerwick	Category 2	Single User

There is also a facility at Scalloway for supply boats to berth, where Northern Offshore Services offer basic agency and other services. The harbour at Scalloway is currently being developed further (primarily for the fishing industry) and it is hoped this development will release more space for general and oil related traffic.

The discoveries which have occurred off the West side of Shetland would suggest that Scalloway might be utilised in the future. Scalloway could perhaps be described as Category 1, Multi User (as could Baltasound, in Unst) but has not been included in the above chart because developments are at an early stage.

Similarly, as part of the Shetland Islands Council policy to share developments through the Islands, it is hoped that some development might proceed at Basta Voe in Yell.*

THE MULTI USER BASES

1 Hudson's Offshore Services, Sandwick

This was the first of the Supply boat bases for which outline planning permission was granted in May 1972 before the major planning decisions on the control of oil related developments had been taken by the Zetland County Council. The old fisheries pier at Sandwick, owned by the Broonies Taing Pier Trust, and the area around it, were developed to provide the base.

Hudsons Offshore became established at Sandwick in January 1973. The company completed Phase One of their development programme, building two berths, warehousing and a quay.

In December 1976, planning permission was granted for a small barytes mill (of approximately 20.000 tonnes/year capacity) to be located on the base at Sandwick. A company, known as Shetland Barytes Limited was formed to operate the mill, being a joint venture between three drilling mud companies — Highland Muds, British Ceca, and International Drilling Fluids — Hudsons Offshore, and a specialised management services company from Aberdeen, Conply. The rock was shipped into Shetland from countries such as Morocco and Spain where the mineral is found, and crushed when demand warranted. It was then exported to the oilfields in powdered form, either directly in supply boats which called at Sandwick or from Lerwick, where it was transported by road to be loaded into supply boats at Lerwick Harbour.

The mill began production in December 1977. Production was sporadic, and eventually in 1979 the mill was sold to the Aberdeen Barytes Company (owned by the Dresser Industries Group) and transported to Lerwick, where it was re-erected on Norscot Base (see below).

See Chart VI "Supply Boat Berths in Operation and with Planning Permission".

OIL SUPPLY BASE FACILITIES IN OPERATION IN SHETLAND (as at January 1981) Chart VII

Facilities	Norscot Oil Services Ltd., Lerwick	Ocean Inchcape (Shetland) Ltd., Lerwick	BP Petroleum Development Ltd., Lerwick	Shell UK Exploration & Production Ltd., Lerwick	Hudsons Offshore Services, Sandwick
Berths Operational	9	3	2	2	2
Water Depth — Mean Low Water Springs	3 berths at 18 ft 6 berths at 21 ft	Minimun 24 ft	18 ft	to 25 ft	21 ft
Pilotage and Navigation Aids	Through auspices of Lerwick Harbour Trust and Pilotage Authority	Through auspices of Lerwick Harbour Trust and Pilotage Authority	Through the auspices of Lerwick Harbour Trust and Pilotage Authority.	Through the auspices of Lerwick Harbour Trust and Pilotage Authority.	On request
Fuel and Fresh Water	Tank farm stocks — one 2500 cu.m gas/oil storage tank; one 4000 cu.m tank; water — 700 tonne capacity. Both fuel and water piped to all berths. Fuel pump rate 80 tonnes/hour; water pump rate 70 tonnes/hour.	Up to 2000 tonnes storage available for bunkering. Spur pipeline for gas/oil to Norscot but buy and sell own fuel. Fresh water available.	No tank farm stocks. Pipeline for gas/oil laid from Norscot in 1977. Fresh water piped to berths.	Pipeline for gas/oil laid from Norscot. 2 x 2500 tonnes oil storage tanks. Fresh water piped to berths. Two water storage tanks.	Tank farm stock — 1400 tonnes fuel oil piped direct to quays; 480 tonnes fresh water storage.
Mud, Cement and Brine	Supplied direct to quayside. Provided by Baroid UK Ltd., B.W. Muds, CeBo UK Ltd., Dresser Magcobar, Halliburton Manufacturing and Services Ltd., International Drilling Fluids, IMCO, Milchem, Dowell Schlumberger	Provided by British Ceca, B.W. Muds, CeBo UK Ltd., Highland Muds and Chemicals, Dresser Magcobar	B.W. Mud operates brine and mixer tanks; CeBo provides bulk chemicals; Milchem bagged chemicals.	Baroid UK Ltd., B.W. Mud, IMCO supply oil based muds from Norscot Base. Ordinary muds provided by CeBo and Dresser Magcobar on base. All other palletised chemicals delivered by trucks.	Available from Lerwick
Warehousing	Seven warehouses, total 17,765 sq metres	Three warehouses, total 45,460 sq ft	One warehouse of 5000 sq ft. New warehouse of 20,000 sq ft due to be constructed mid 1981, completed mid 1982.	One warehouse of 3500 sq ft	Two warehouses, each 27,000 sq ft plus one insulated (heated) warehouse 70 X 38 ft suitable for storage of perishable items or use as a workshop.
Mechanical Handling	2 x 110 ton cranes, 1 x 60 ton crawler crane, 3 x 35 ton cranes, forklift up to 10 ton capacity, articulated wagons for pipes. Normal and specialist pipe trailers. Range of vans, pick-ups, normal flatbeds, etc.	60 ton crane, 50 ton crane, 2 x 36 ton cranes, 27 ton cranes, three forklifts, seven articulated vehicles, trailers	36 ton crane; 5 ton forklift.	2 x 35 ton rough terrain cranes, 1 x 50 ton rough terrain crane, four articulated tractor units, seven trailers exclusively for Shell's use for pipeyard, 60 ton crane, two forklifts on base.	1 x 30 ton crane for boat loading and stock yard work; 1 x 5 ton forklift. Trailers for pipe movements.
Office Accommodation	2 x 800 sq. metres office accommodation in two office blocks; 300 sq. metres offices attached to warehouses	Six offices available for rent, each approx. 220 sq. ft., 1320 sq. ft. total	BP only	Only for Shell	14 offices, kitchens and washrooms
Crew Changes and Accommodation	Helipad rated up to Sikorsky S61 N, including diversions. Agency can make arrangements for crew changes through Lerwick, Sumburgh or Aberdeen	Agency Service provided	None	Hay & Co. act as agents at Sumburgh Airport	Two airport offices at Sumburgh to service rig crew changes by helicopter; also handle customs clearance on fixed wing aircraft for personnel and materials. In event of crews stranded overnight accommodation arranged

OIL SUPPLY BASE FACILITIES IN OPERATION IN SHETLAND (as at January 1981) — Continued

Facilities	Norscot Oil Services Ltd., Lerwick	Ocean Inchcape (Shetland) Ltd., Lerwick	BP Petroleum Development Ltd., Lerwick	Shell UK Exploration & Production Ltd., Lerwick	Hudsons Offshore Services, Sandwick
Telecommunications	PABX 7 switchboard, telex machines, VHF radio, telecopiers, comprehensive internal telephone system, tone paging system (see text)	Telex and telephones, UHF radio	Telex, telephones	Telex, telecopier, direct line to Shell in Aberdeen, VHF private channel 29 x 100, six outside telephone lines. Telephone link to platforms, Norscot tone paging (bleeper) service. Radio link from office to pipeyard, base and cranes.	Full VHF service available for rig and supply boat communications. Telex, telephones
Engineering Facilities	Comprehensive engineering service including machinery for medium and large fabrication in steel, aluminium, etc. Oil tool work, i.e. refurbishing tubulars. Total workshop area 2 x 540 sq. m. Welding hall has facilities for semi-automatic and manual welding. Paint and blast shop 315 sq. m.	Comprehensive engineering service. 10,000 sq. ft. workshop with overhead cranes. 2nd oilfield lathe capable of turning flanges up to 53 inch diameter. Two lathes up to 20 inch diameter for cutting; profile burner, machine to make machine tools	Facilities around Lerwick Harbour utilised	Contracted out to various machine shops. Since April 1980 excess and recovered pipe from field returns for inspection and repair	Not available
Shipping and Forwarding	Full range of agency services provided, including Customs documentation. HM Customs office on base — mainly used for helipad	Full range of agency services provided. Agents for Ocean Inchcape Limited Supply ships	Hay & Co. are agents	Customs documentation for own materials	Sea and air charter and freight services, expediting and customs documentation at all Shetland points of entry
Area	42.871 acres total site area	5½ acres open storage at Point of Scatland; 2.05 acres at Gremista pipeyard. Backup land of 12 acres available with planning permission	2¼ acres; additional pipe storage is being prepared	10 acres for pipe storage plus two acres at Ocean Inchcape base on contract, and emergency pipe storage facility of up to two acres at Norscot base	10 acres for open storage
Ancillary Services	Container hire; purchasing stocks carried for sale; base client has cold store and comprehensive victualling service. BOC provide a service for gases	Container hire; base client has cold store, and comprehensive victualling service	Small amount of local purchasing	Only local purchasing. Ships do their own victualling, 10 acre casing yards operated by contractor	Dockers can be arranged; personal travel arrangements can be organised

NOTE: At Scalloway, where there is one berth operational, Northern Offshore Services provide a full ships agency service on request: pilotage, limited fresh water and fuel, mud and cement and dock labour and craneage.

Facilities include: 24 hour ships agency and customs documentation service, VHF radio channel 16, radar and Decca.

In March 1980, BP Minerals bought Selection Trust which owned Hudsons Offshore. It is hoped that some use will be made of the good open storage warehousing, and office facilities that the base has and that some development might be stimulated there.*

2 Norscot Oil Services Limited, Lerwick

In March 1973, the Local Planning Authority granted permission for a supply base at the Greenhead, Lerwick, to Norscot Oil Services Limited a company wholly owned by Fred Olsen Limited of London (see organigram of company structure). A programme of land reclamation was instituted and completed by 1975, which resulted in a sizeable area of flat land created, to provide quays and other facilities. The land on which the base is situated is leased from Lerwick Harbour Trust, encompassing approximately 43 acres. The Norscot Base is the largest supply base** in Shetland, both in terms of land area, facilities offered,*** and staff employed. The base has gradually evolved and represents an investment considerably in excess of £5 million.

Apart from providing the basic facilities to allow mooring of supply boats, and supply of mud, cement and brine stocks, fuel oil bunkers and fresh water*** the base has developed comprehensive materials handling, storage, and engineering facilities.

The Engineering Facilities

The engineering aspect of the Norscot base began to develop in 1976; the years 1977/78, reflected the movement offshore Shetland from the exploration to the development phase of some of the major oilfields and fabrication work on a considerable scale began to take place at the base.

For example, in 1977 a major completion contract worth in excess of £1 million (1977 money) was undertaken on two Chevron modules, which were subsequently installed on the South Ninian platform. This work lasted three months, with the accommodation ship 'Christian IV' being brought into Lerwick Harbour and moored at the Base, in order to house the additional workforce of 150 men.

Whilst having a core of engineers working full time at the base, Norscot Services, as part of a major company structure, is able to call upon specialist labour services at short notice in order to undertake such contracts (see organigram of company structure in Appendix).

Major contracts for offshore operators that the engineering workshops have fulfilled, include —

(i) **Brent System (Shell Expro) and Thistle 'A' (BNOC)*****

Complete fabrication of numerous pipeline tie-in spool pieces from 16 to 36 inches in diameter and up to 400 ft. in length;

(ii) **Ninian System (Chevron)******

Welding of connectors;

(iii) **Brent 'A' 'B' (Shell Expro) and Heather 'A' Fields (Union Oil)**

Module outfitting which included all trades and totalled 70,000 manhours;

(iv) **Ninian Central Platform (Chevron)**

Completion and outfitting work on two gas separation modules which included all trades and totalled over 120,000 manhours;

(v) **Thistle 'A' Platform (BNOC)**

Mechanical maintenance and construction contractor since February, 1978 (in conjunction with associate company Blandford Offshore Services Limited).

Facilities offered by the base can be seen in Chart VII.
**See diagram of base layout.*
***For more detail listing facilities see Chart VII "Oil Supply Base Facilities Operational in Shetland (as at January 1981)".*
****See map showing Brent System pipeline links from the oilfields to Sullom Voe in Chapter five.*
*****See map showing Ninian system pipelines in Chapter five.*

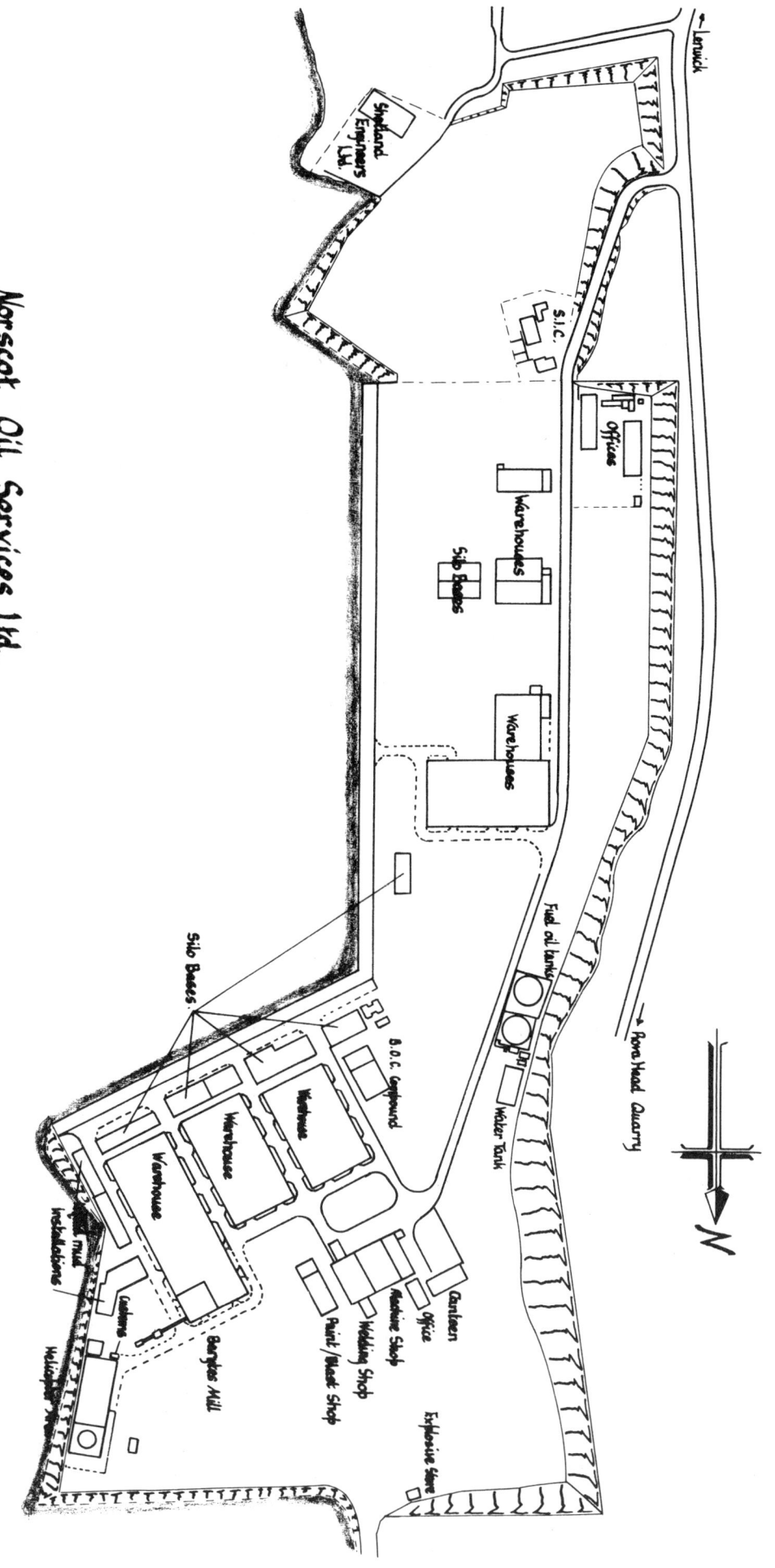

Norscot base has also carried out a considerable amount of engineering work to support onshore developments. For example, for the Sullom Voe Terminal, Norscot did the mechanical outfitting work on two Pre Assembled Units (PAU's — see Glossary) each weighing 400 tons, which were offloaded at the base from a heavy lift vessel using modular bogies. This work lasted 50,000 manhours. Other onshore work has included the installation of liquid mud plants on the base for Dresser Magcobar and Milchem (in 1979). The tanks and pipework for Milchem were fabricated on the base. Other fabrication, of large spool pieces, pipe bridges, hoppers and containers, has been carried out on the base, as well as ship repair. The range of jobs done is wide — even including the steelwork for the new school built at Brae, Shetland.

Material Handling and Storage Facilities

Materials handling and storage services were amongst the first to be developed at the Norscot base. For example, in 1975 and 1976, Norscot handled (for BP) 175,000 tonnes of concrete coated pipes for the 152 kilometre long Ninian pipeline. The pipes were shipped from Invergordon to Lerwick by Norscot then handled, stored and loaded out into the client's pipe carrying and supply vessels. Norscot also has a shipping and materials handling contract from Shell/Expro for the Brent pipeline maintenance, and about one third of the original supply of the Brent pipeline was stored and handled through the base.

Another long term contract for Norscot has been to act as the onshore supply base to support construction and production operations on the Ninian Field, for Chevron. This has included the storage and load out of 70,000 tonnes of casing plus associated drilling muds.

Norscot Oil Services Limited also contract out labour to other companies. For example, from July 1976 the base has provided materials handling and shipping services for all the Terminal construction materials at Occidental's Oil Terminal at Flotta in Orkney and continues to do so. Initially men from Shetland travelled to work there for periods of several weeks at a time, but now they employ Orcadians under Norscot management. Norscot also has an ongoing materials handling contract with BP to provide labour and plant to operate their supply base at Lerwick.

Other Developments

(i) **Containerisation**

In the middle of 1978 the move to containerise supply boat cargoes was reflected in Norscot becoming agents for E.P.D. and the Container Trading Company (U.K.) Ltd. of Aberdeen. Mini containers for offshore use are stored at the base. Dry freight, refrigerated, open topped, flat, and other containers, built to Lloyds specifications, are available through these companies. The offshore container business is highly specialised, the containers being specially strengthened for safe heavy lifting and subject to frequent inspection and maintenance.

(ii) **Communications — Paging Service**

The North Sea oil supply business requires people to be available on a 24 hour basis. As a result of this, Norscot has installed a paging system, using UHF radio pagers manufactured by Motorola. A special transmitter aerial was erected on the island of Bressay (see Lerwick Harbour map) to give the system the widest possible range, with the control centre located adjacent to the main switchboard and night service. This service is now widely used, and is available not only to base and oil industry personnel, but to all those in the community who require it.

(iii) **Barytes Mill**

In 1979, a barytes mill was established at Norscot Base by the Aberdeen Barytes Company. The crushing and milling machinery was obtained from Hudsons Base at Sandwick (15 miles south of Lerwick) when the Shetland Barytes Company, who were operating this mill, sold out to the Aberdeen Barytes Company, which is owned by Dresser Magcobar. The barytes ore comes from Foynes in Southern Ireland in 2500 tonne shipments, and a stockpile is maintained on the base. The mill is kept constantly in operation to meet offshore needs.

At January 1981, there were 24 client companies which had contracted services such as offices and warehouse accommodation, labour, and equipment for the handling of materials at Norscot. These companies provide a variety of oilfield materials and services such as drilling mud, and brine, non destructive testing, and electronic care and maintenance to the offshore industry.*

See Appendix to Chapter Three for Listing of Norscot Base clients.

Together with the base facility of materials handling, storage, port facilities, and a modern well equipped engineering machine shop, fabrication shop, and paint/blast shop, the future business known to Norscot at this time, includes the fabrication of various tie-in spool pieces for offshore subsea connections, the rethreading and maintenance of all sizes of casing for most of the major oil companies. Norscot will also handle about one third of the Magnus pipeline and associated onshore support for the Saipem Castoro Sei pipelaying barge. This is in addition to the ongoing contracts for the materials handling for BP in Lerwick, Occidental at Flotta and for general engineering work and fabrication at Sullom Voe.

3 Ocean Inchcape (Shetland) Limited (OIL)

Since 1973 Ocean Inchcape Limited have provided a range of oilfield support services from Lerwick. Ocean Inchcape were first established at Lerwick Harbour in premises rented from Fraser and Partners (now amalgamated with Barratt Construction (Aberdeen) Limited) and Hay & Company.* All the plant that the base used was owned by the base. OIL also had a 2.5 acre pipeyard on ground rented from Lerwick Harbour Trust (which they still have), and utilised the facilities of Lerwick Harbour for their shipping operations.

A radical change occurred in the mode of operation of the Company, when Ocean Inchcape decided to implement the outline planning permission obtained in June 1975 from the Shetland Islands Council, to construct a purpose built supply base at the Point of Scattland in Lerwick Harbour. The Company finally moved out of its premises in the North Ness in May 1978, and had the new £2.5 million base at the Point of Scattland ready for formal commissioning on 25th August 1978.

Ocean Inchcape's new base has allowed the company to expand and develop its business activities. Like the other bases at Lerwick Harbour, Ocean Inchcape provides storage for drilling muds and chemicals and for casing, having five mud companies with warehousing and silos on the base, and five oil companies using the open storage facilities.**

Containerisation of drilling muds and chemicals has caused Ocean Inchcape to develop its container hire business. The company has its own pool of mini containers, used mostly to transport drilling mud materials offshore, and on hire to Amoco, Chevron, Conoco and Shell. In this context a container inspection and maintenance service is provided by the base engineering staff to meet Lloyds requirements.

Engineering Facilities

A variety of engineering services have been developed at the base, covering both vehicle and plant repair and maintenance and fabrication work.

In respect of vehicle maintenance, Ocean Inchcape are authorised Lucas Keinzle tachograph agents, and provide a tachograph station for the Islands. The company also do vehicle and plant repairs and maintenance not only to meet the needs of the base but also for companies including LHD and the North of Scotland Hydro Board (these two organisations are based at Lerwick Harbour).

Engineering repair and maintenance is also carried out on shipping. For example, Ocean Inchcape repair and maintain the tugs at Sullom Voe for Shetland Towage Limited (a company also, like Ocean Inchcape, affiliated to the Ocean Transport and Trading Group).

Fabrication maintenance has also been carried out on car ferries owned by Shetland Islands Council and on a barge owned by Nordic Offshore Services.

Other engineering work the Company has undertaken includes work on spool pieces, maintenance fabrication and pipework on Lerwick Power Station, and work on fuel tanks.

Ocean Inchcape were in joint venture with Hay & Company of Lerwick — the resulting company being Ocean Inchcape (Shetland) Limited. This joint venture ceased once Ocean Inchcape became established at their new base at Point of Scattland, and the company is now wholly owned by Ocean Inchcape Limited.

**See Appendix to Chapter Three for listing of OIL Base clients.*

Layout of Ocean Inchcape Limited Service Base: Point of Scattland: Lerwick

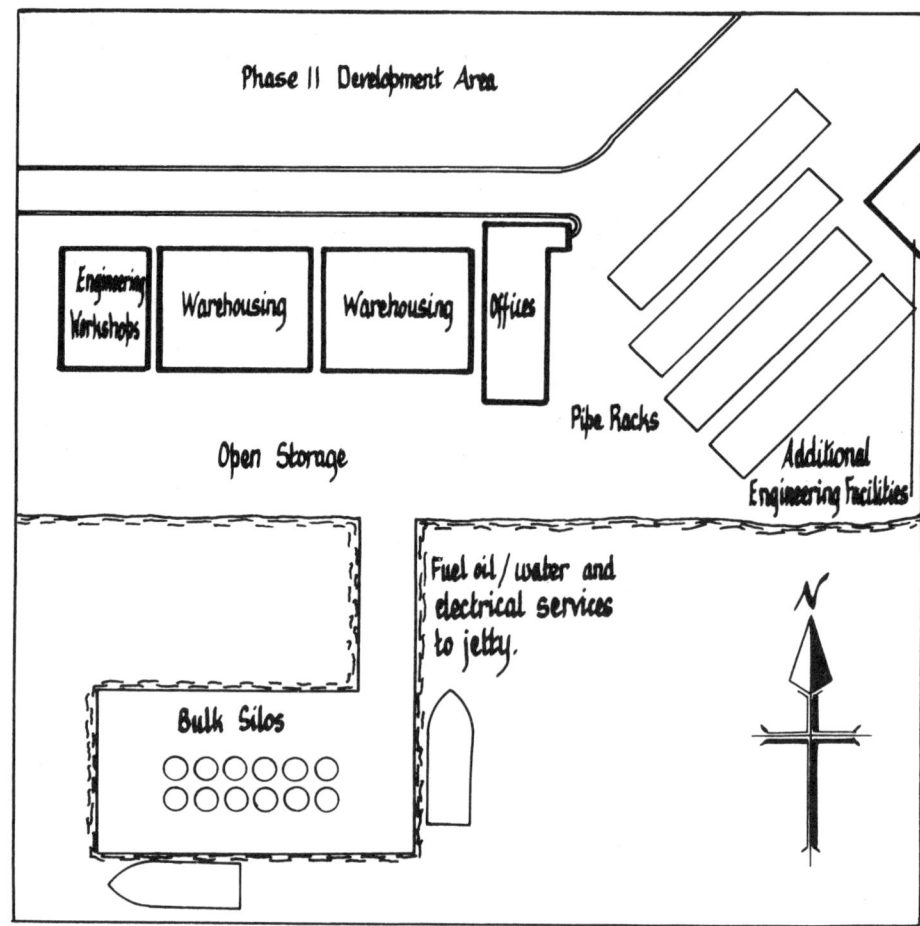

The work of the base continued to develop and expand in 1980, its proximity to the Shell Base proving useful in helping to meet Shell's requirements for extra berthing and supply boat facilities.* Expansion also occurred in respect of services to the other oil companies and with more oilfield developments imminent offshore Shetland, Ocean Inchcape hope to develop an additional 12 acres, for which they have planning permission, to provide more open storage, warehousing and offices.

THE SINGLE USER BASES

1. BP Petroleum Development Limited, Holmsgarth, Lerwick

BP obtained planning permission in May 1974 and work began in late 1974 to construct this purpose designed supply base. Much of the Lerwick Harbour Trust land on which the base now stands was reclaimed from the sea. The base is only in business to meet BP's supply needs, acting as a forward base for BP's main North Sea supply base in Dundee.

In 1980 BP had five service vessels on permanent charter — two supply vessels for Forties platforms, with three anchor handling/supply vessels for mobile rigs, two of which were North of 59 degrees North. The principle the company works on is that if there is a mobile rig off Shetland, an anchor handling vessel is always relatively close at hand, and this vessel would not leave unless there was another vessel to replace it. For the deeper waters off Shetland a higher horse powered vessel is required as mooring becomes heavier. Such vessels might have only 1,000 tonnes maximum carriage capacity for a combination of bulk and deck cargoes.

In 1981, the Company was awarded a further contract by Shell to continue and develop its services at Shell's pipeyard in Lerwick.

The platform supply situation is somewhat different and BP are anticipating expansion with the development of the Magnus Field. There will be at least two additional vessels chartered for the construction period on the Magnus Field.

The mode of operation of BP's service vessels has changed since the early 1970's. As with other operators, the number of supply boats per drilling rig has been gradually reduced from 3 to 2½ to 1⅓. This is due, amongst other factors, to the increased size of the vessels used, reduction in idle time through more sophisticated scheduling arrangements and the principle of "Multidrops" where one vessel makes several deliveries to different locations during one voyage.

Whilst BP charters anchor handling and supply vessels, the company prefers to own more sophisticated vessels such as the £9.4 million safety support vessel (SSV) for the Magnus Field which will provide oilfield support services in respect of fire fighting and rescue. It also owns composite craft such as the Sulair which, with a 65 metre clear cargo deck length is a most useful platform supply vessel in winter, converting to diving support in the Spring and Summer, when its dynamic positioning facility and moonpool are utilised. These vessels are manned by BP's own marine crews as well as the Forties Kiwi which supports the Forties Field with a fire fighting, rescue, maintenance support and limited accommodation.

At Lerwick BP is intending to construct more warehousing — approximately another 20,000 square feet to add to the original 5,000 square foot warehouse, with a view to a mid 1982 completion date. BP has also come to an agreement with Lerwick Harbour Trust in respect of the use of another berth for its supply operations, adjacent to the Holmsgarth base. There is also an additional 2½ acre casing yard being prepared to meet future requirements.

This will mean some more jobs being created locally. People will be employed not by BP directly but by Norscot Oil Services Limited who at present have a three year labour contract to service BP's base operations. There will remain only one BP employee on the base — the Materials Controller.

A continual expansion through the 1980's is envisaged by BP, of its operations in Shetland. For the latter half of the 1980's the West side of Shetland is expected to be developed including the Clair Field. On the East side BP also have their discovery near the Ninian Field to evaluate, as well as a programme of further exploration.

As one would expect with a forward base operation it has been marine bulk items such as mud, chemicals, fuel, and tubulars which have been supplied from this base. Specialised drilling tool repair services operated by local companies, and other services, have sometimes been used when supply boats bring in damaged equipment owned by BP. It is hoped to expand such facilities at Lerwick to cater for the expected increased supply boat usage.

---

*See Chart VIII "The Pattern of BP's Lerwick Base Activity 1976-1980".

THE PATTERN OF BP's LERWICK BASE ACTIVITY 1976 - 1980*

Chart VIII

BP SUPPLY BOAT CARGOES 1976 - 80
(metric tonnes)

Year	Lerwick	Dundee	Gt. Yarmouth	Total	% of total through Lerwick
1976	25268	392117	9928	427313	6
1977	58820	365544	6287	430651	14
1978	64630	397508	6652	468790	14
1979	51314	324815	7218	383347	13
1980**	137639	283839	5936	427414	32

FOOTNOTES:

*Figures not available before 1976

**A dramatic increase in tonnages and the percentage total of cargo through Lerwick can be seen in 1980. This trend is expected to persist with further developments imminent. The average tonnage carried per voyage from Lerwick rose from 221 to 634 tonnes from 1979 to 1980. Main increases being in bulk chemicals, gas oil, and water.

BP SUPPLY BOAT VOYAGES 1976 - 80

Year	Lerwick	Dundee	Gt. Yarmouth	Total Voyages	% of total through Lerwick
1976	159	441	85	685	23
1977	262	344	73	679	39
1978	288	404	59	751	38
1979	232	366	74	672	35
1980	217	288	57	562	39

NOTE: BP operates a co-ordinated marine supply effort from its three centres at Lerwick, Dundee and Great Yarmouth.

2. Shell UK Exploration and Production Limited

The first offshore marine supply operation out of Shetland started on 26th May 1971 when Shell UK Exploration and Production began operating from Lerwick Harbour. Shell began operations from Victoria Quay and then from Ocean Inchcape's former facility at the North Ness. Shell then decided to have their own purpose built base in Lerwick Harbour.

Planning permission for this base was obtained in May 1974, and it began to function in 1974. The land on which it is sited, like the adjacent land at the Holmsgarth site on which the BP base is situated, has been to a great extent reclaimed from the sea, and is owned by Lerwick Harbour Trust.

The base acts as part of an integrated supply operation in conjunction with Shell's main supply base at Aberdeen and other ports Shell use on the Scottish Mainland. A measure of activity could perhaps be seen from the following figures:

Shell UK Supply Vessel Sailings

Ports:	Aberdeen	Lerwick	Montrose	Peterhead	% Total from Lerwick
1979	602	439	110	87	35.4%

Note: If one was to compare Lerwick and Aberdeen sailings only, in 1979, Lerwick would account for 42% of the total.

A further measure of activity is the cargo throughput of the Shell base, which was greater than that of any of the other bases in Lerwick Harbour in 1980, at over 230,000 tonnes. Approximately 27 per cent of this tonnage was imported through the Shell base, and materials such as casing, mud, chemicals and fuel were imported at other points in the Harbour and transported to Shell base for export.

One cannot be surprised at a throughput averaging over 19.000 tonnes, from 42 sailings per month, when one considers the complexity of developments that Shell is supporting offshore Shetland — namely three oilfields — Brent, Cormorant and Dunlin — and six platforms, plus all ancillary craft engaged in diving support, maintenance, hook-up and pipelaying.

Shell has 12 or more supply boats (depending on season) on charter to support its Northern North Sea Operations — most of these boats being active in supplying the area offshore Shetland.* Changes have occurred in the mode of operations. At one stage Shell had 24 supply vessels on charter; for those they now have, to cover seven platforms and three exploration rigs** (i.e. over one per vehicle average), there has been no reduction in the tonnages being transported, merely in the number of ships. Larger, more versatile ships are now used. For example, before 1975 some vessels were exclusively anchor handling; since then there has emerged a second generation of vessels able to handle anchors and also carry cargo for platforms.

Scheduling has also become more complex, with multi-destination loads carried by the ships, as they make their "milk runs".

Presently two vessels are dedicated to Lerwick. Lerwick acts principally as a source of bulk supplies for tubulars, casing, chemicals, mud, cement, gas oil and water. All deck cargo leaving Lerwick is discharged by the supply boats; the under deck cargo (in tanks of barytes, bentonite, cement and water) is kept as floating storage, to be ready on demand. A quicker turn around time is possible from Lerwick, perhaps also reducing the number of ships required here, as it is only eight hours sailing time to the Brent Field complex off Shetland, compared to 22 hours from Aberdeen.

Although there are only seven staff directly employed by Shell on the base, there are 40 or more people (in total) involved in its operations and working there under various contracts. The biggest of these contracts is for the operation of Shell's ten acre pipeyard at Gremista, which employs 26 men, under a contract with the Ocean Inchcape base.

See Diagram 2 "Brent Field — Marine Activity 13th July 1978" as example of activity offshore Shetland.

**Not all of these are located offshore Shetland.*

Diagram II

Brent Field - Marine Activity 13th July 1978

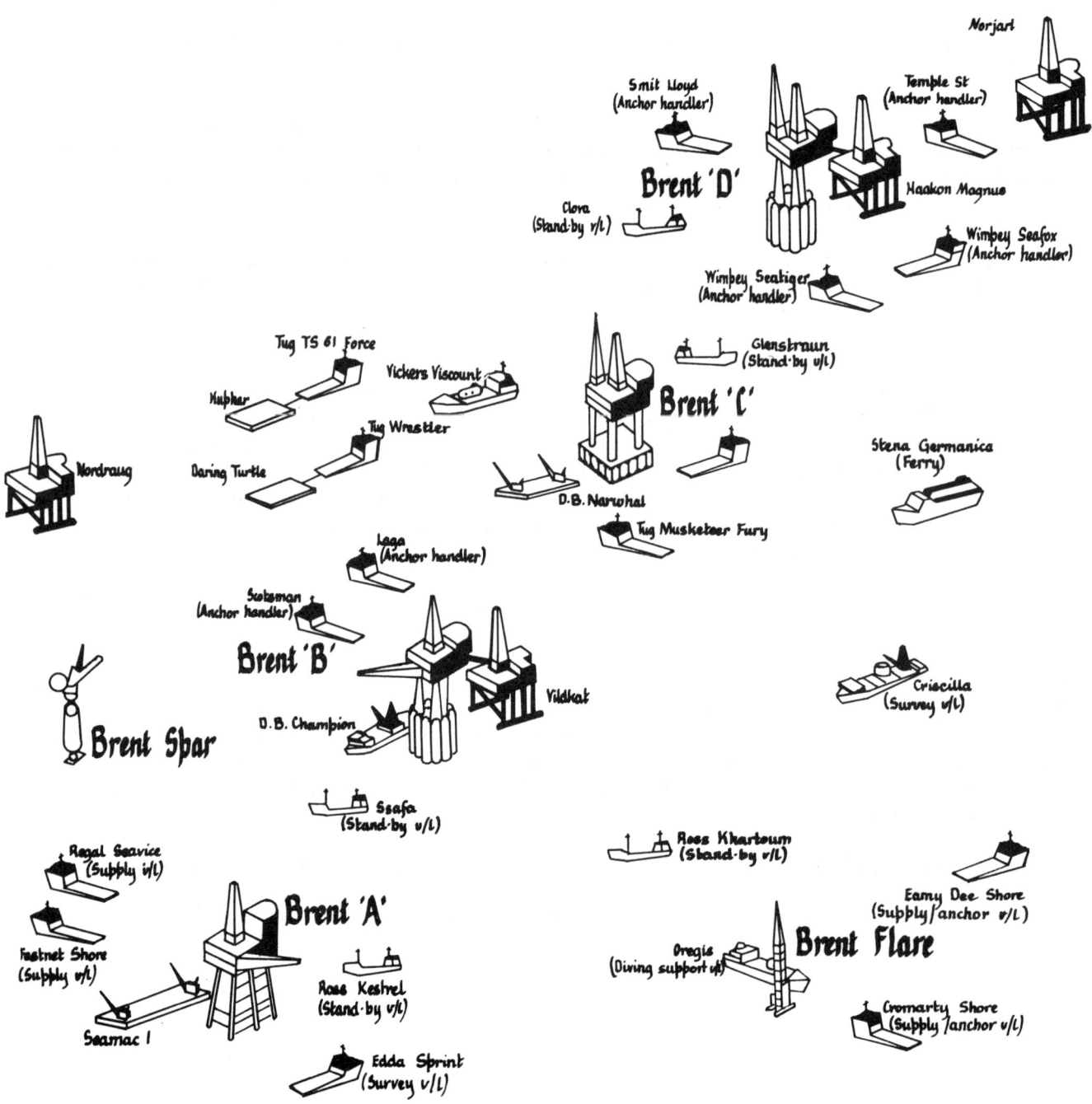

The Drilling Muds and Chemicals Business

Having examined the oil related supply business in Shetland and the supply bases, it is evident from a study of the oil related cargoes that drilling muds and chemicals are of major importance.*

A total of no less than 11 companies (nine currently active) plus a barytes mill have a presence at the various bases around Lerwick Harbour in order to engage in this business, which has been estimated to be worth over U.S. $30 million per annum.

The drilling muds (or fluids) and chemicals business caters to the requirements of the oil companies drilling wells offshore Shetland. When a well is drilled it is essential to have fluid circulating in order to cool and lubricate the drill-bit, remove rock chips and shavings and drill cuttings from the point of drilling, and to control the pressure in the well and prevent blow-outs.

Some changes have occurred, since the first "mud" companies arrived in Shetland in 1972, in terms of the materials being handled. Originally, it was mainly the powdered form of the mineral barytes, which was exported in bags or in bulk to the rigs. This mineral was then mixed with ordinary water or seawater and perhaps some other chemicals to provide a drilling fluid for the wells being drilled, mainly acting as a weighting agent to prevent the bottom of the well collapsing. Once the well was drilled, this mud would then be discarded.

Since about 1977 liquid mud has been exported in increasing quantities from Shetland. Liquid mud usually consists of approximately 60 per cent marine gas oil, 20 per cent water and 20 per cent solids. The solids will include barytes or other chemicals for weight, and perhaps also lime to control the acidity content, gelling agents and emulsifiers to prevent the solids separating out of the solution.**

The usage of liquid mud depends to some extent on the geological structure being drilled. If, for example one has a very porous rock, such as swelling shales, there can be a tremendous loss of ordinary water based drilling fluid into the rock and this loss of fluid circulating would seriously impede drilling. Using oil based drilling fluid prevents this occurring to the same extent.

Liquid mud is usually mixed in Shetland and exported in that form to the drilling rigs and platforms in supply boat tanks. This eases offshore storage and handling problems which result from having to mix it offshore.

Liquid mud is also useful because it can be re-used. It is returned to Shetland where it is "cleaned" of rock chippings and other wastes, by using centrifuges and sieving. If needed the specific gravity can be altered by adding diesel or centrifuging out the barytes to lighten the liquid, or adding more barytes or other chemicals if a heavier mud is required next time round. The mud can, however, be reconditioned to the original specifications should the oil company wish to continue using mud of the same specific gravity and chemical content.

The average cost of liquid mud is from £30 to £40 per barrel*** depending on the weight of the mud. The heavier the mud,**** the higher the price as a greater barytes content will be needed and therefore more gelling agents to hold the barytes in suspension. The weight is measured in pounds per gallon (lb/gal.) or pounds per square inch per 1000 feet (psi/1000 ft.) to take account of well pressures.

Another important development has been the increased usage of brine offshore. Brine is used because it has been discovered that when a well is capped and is full of seawater, the plankton will breed in the seawater and create oxygen. Over a period of time there will be a build up of gas and this can cause the plug on the well to blow. There are different types of brine, the most common being mixed from potassium chloride, calcium chloride and occasionally calcium bromide to add weight to the liquid. This high concentration of chemicals kills the bacteria in the seawater so that there is no build up of gas in the well plugged.

---

*See Appendix to Chapter Three for detailed listing of Lerwick Harbour's Import and Export statistics.

**Bentonite is one of the minerals widely used to keep solids in suspension.

***Of the major constituents of liquid mud:— Diesel costs £170 per metric tonne, barytes from £37 to £50 per metric tonne (1980 money).

****For a really heavy mud, the mineral ilmanite is sometimes used in preference to barytes as it has a very high specific gravity.

Mud Company Operations in Shetland

Four of the nine companies currently active in Shetland* are related to each other, namely Baroid and CeBo; Halliburton and Imco. These related companies work in collusion rather than in competition, as each has a specific function, for example, Baroid has specialised in liquid mud, CeBo in powdered barytes and chemicals. Even within the other five companies there are those who have a particular speciality — for example, Dowell Schlumberger provides cement to secure the casing and sides of the well-holes drilled — and thus have their own corner of the market. Other arrangements also exist between the companies. For example, CeBo acts as the wholesaler for some of the others as CeBo can import barytes and bulk chemicals in their own specialised ships from Holland at £8 per tonne carriage, whereas it would cost £50 per tonne if these materials were shipped on the normal shipping service from Aberdeen. The oil companies operate in a combination of ways — for major field developments, mud companies can get contracts to supply mud for a two year period. Thereafter the contract is reviewed, and usually open to re-tendering by the various companies. This allows the oil companies to obtain a better price for their materials. For some exploration wells, and in some cases, fields in production, some oil companies go to the spot market. In this case negotiations and counter bidding would take place along the waterfront, until the oil company is satisfied it has achieved the lowest possible price. This combination of methods helps to share out the business, which, as stated, is valued in many millions of dollars per annum.

The investment in silos and tanks alone in the Lerwick harbour area, to provide these materials, is probably in excess of £1 million. For example, the cement base for one 220 tonne silo, the silo itself, and its metering and pumping instruments can cost in excess of £25,000. The provision of muds, and chemicals to support offshore drilling needs can thus be seen to be big business, and very essential to the marine supply activities operating from Shetland.

Lerwick Harbour

Having described supply base developments, it will be noted that four of the five supply bases in Shetland are in Lerwick Harbour. The supply boats calling at these bases use the pilotage and other facilities provided by the managers and owners of the Harbour, Lerwick Harbour Trust.

Lerwick Harbour Trust

The development of the port of Lerwick would have been inconceivable without the careful direction and management of the Lerwick Harbour Trust. Lerwick Harbour Trust was formed by the Lerwick Harbour Improvement Act of 1877 "for the maintenance and management of the Port and Harbour of Lerwick and the shipping thereto".

In 1952 an Act of Parliament authorised a major scheme to enlarge the quays and wharves and also to extend the harbour limits. These works were carried out between 1955 and 1960. Since 1971, and the growing pressure of oil related developments, further legislation has been enacted each year to extend the borrowing powers of the Trust, to increase their area of jurisdiction and to construct new works. The Act of 1973 gave the Trust powers to borrow up to £3 million and this was raised to £5 million by the 1974 Act.

The Trustees who govern Lerwick Harbour include the following:—

(i) Three members elected from eligible shipowners;

(ii) three members elected from eligible ratepayers;

(iii) six members appointed by the Shetland Islands Council from Council members, of whom three are representatives of Lerwick wards.

The Trustees meet at least once a month, while the various committees meet as often as required.

In 1970, the day to day running of the port was carried out by the Clerk, harbourmaster and eight employees. Today, the staff includes the Clerk and General Manager, his deputy, the harbourmaster, assistant harbourmaster and 45 employees including office staff, berthing masters and pilots.** It can thus be surmised

---

*See Chart IX "Companies Supplying Barytes Cement and Chemicals from Shetland Offshore (as at January 1981)".
**See Chart X "Oil Related Employment Created by Maritime Supply Needs".

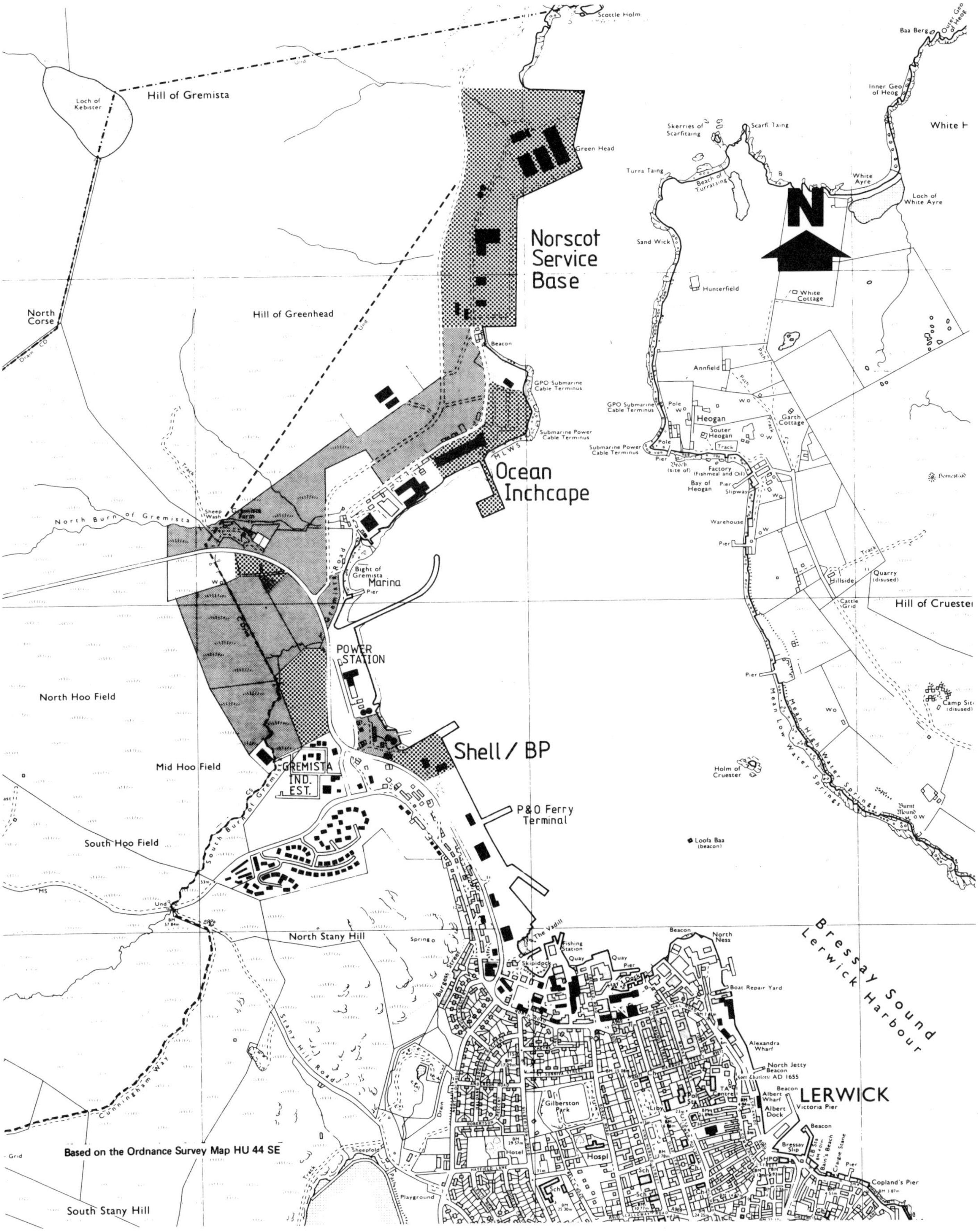

Oil Related Developments at Lerwick Harbour

Existing

Proposed

COMPANIES SUPPLYING BARYTES, CEMENT AND CHEMICALS FROM SHETLAND TO OIL COMPANIES DRILLING OFFSHORE (as at January 1981)   Chart IX

Company	Shetland Location	UK Head Office[*1]	Parent Company	Materials Provided
N L Baroid UK Ltd.	Norscot Base	London	Houston	Drilling fluids and chemicals
British Ceca	Ocean Inchcape Base, Lerwick	London	Paris	Drilling
B W Muds Limited	Norscot Base, Lerwick	London	Was in USA but now changed to UK	Oil based mud, brine and dry chemicals
CeBo	Lerwick	Aberdeen	Netherlands/ USA	Cement, chemicals, barytes, bentonite
Dowell Schlumberger	Norscot Base, Lerwick	Aberdeen	Houston	Bulk cement, chemicals and additives
Dresser Magcobar	Norscot Base, Lerwick	London	Houston	Drilling mud and chemicals
Halliburton M & S Ltd.	Norscot Base, Lerwick	London	Houston	Chemicals, cement, and cement additives
Highland Muds & Chemicals Ltd.	Ocean Inchcape Base, Lerwick	Doncaster, Yorkshire	UK	Drilling fluid, chemicals
Imco Services (UK) Ltd.	Norscot Base, Lerwick	London	Houston	Drilling mud and chemicals
International Drilling Fluids	Norscot Base, Lerwick	Uxbridge, Middlesex	Calgary, Alberta	Drilling muds and chemicals
Milchem Drilling Fluids Ltd.	Norscot Base, Lerwick	London	Houston	Chemicals

TOTAL NUMBER OF COMPANIES — 11
*FOOTNOTE: *1 Although many of the companies have UK head offices in London, most of the Shetland Offices report to area offices in Aberdeen as there is a co-ordinated supply effort from Aberdeen and Lerwick to service the area offshore Shetland.*

Facilities — No. & Capacity of Silos	Date of Arrival in Shetland[*2]	No. of Employees in Shetland in 1980	Comments
11 liquid mud tanks	1975	3	Baroid have a liquid mud cleaning facility — including a centrifuge, and double deck shaker to screen drill cuttings out of the mud.
2 silos; 2000 sq ft warehousing			
warehousing at Norscot and Ocean Inchcape; 1 silo, 18 tanks, 11000 bbls capacity	1977	4	Brine plant at BP Holmsgarth base. Plant expanding. Brine also supplied to Brent A B C Platforms and Cormorant A.
25 silos, 4-4500 tonnes at Norscot, Ocean Inchcape and Shell bases and various locations; 18000 sq ft warehousing at Norscot	1972	7	CeBo obtain bulk supplies from their mill in Ijmuiden in Holland. As well as supplying direct to offshore they also wholesale barytes and other items to other mud companies. In addition their fleet of vehicles is used to transport cement, helping in all major construction projects which have occurred in Shetland.
5 silos — 900 tonnes capacity; Chevron warehousing	1972	2	Only bulk oilwell cement used. Specialists in oilwell cementing; started 1972 at Sandwick; moved to Lerwick in 1977.
1 silos, 2680 tonnes capacity; 9 x 600 bbls tanks liquid mud; 3 warehouses totalling 27550 sq ft	1972	4	Dresser obtain barytes supplies from Aberdeen Barytes Company mill (owned by a parent company) at Norscot Base. Approximately 20,000 tonnes milled for Dresser at Norscot Base in 1980.
1 silo x 120 tonnes; 5000 sq ft warehousing	1975	1	Halliburton own Imco Services, and share a warehouse with them. Through their Aberdeen office cementing crews are provided offshore.
1 silo; 4000 sq ft warehousing			
12 silos: 10 x 500 bbls; 2 x 30 metric tonnes; all at Norscot Base warehouse (see Halliburton)	1976	2	Imco have oil based mud plant and a centrifuge to clean and refurbish mud returned from offshore, for re-export.
No silos owned by IDF. Warehousing at Norscot	1976	1	Mainly bagged chemicals from Shetland; most business ex Aberdeen.
3 silos: 2 x 50 tonnes; 1 x 30 tonnes; Warehousing 7000 sq ft	1978	1	Mainly chemicals, starches, polymers, detergents; some barytes.

[*2] *Most companies moved into Shetland when the production phase began offshore (from 1975)*

OIL RELATED EMPLOYMENT CREATED BY MARITIME SUPPLY NEEDS — Chart X

Location in Shetland	Employing Organisation	1972	1973	1974	1975	1976	1977	1978/79 1980
Sandwick	Hudsons Offshore	-	7	12	12	10	7[*1]	2
Lerwick	BP Petroleum Development Ltd.	-	-	1	1	1	1[*2]	1
Lerwick	Norscot Oil Services Ltd.	-	2	7	66	99	241[*3]	200
Lerwick	Ocean Inchcape (Shetland) Ltd. (O.I.L.)	N/A	N/A	N/A	12	31	33	63
Lerwick	Shell UK Exploration & Production Ltd.	3	4	4	6	6	7	7
Lerwick	Lerwick Harbour Trust	18	28	33	38	38	43[*4]	45
Total for Lerwick Harbour Area		21	34	41	123	175	325	316
Total for all areas (i.e. Lerwick + Sandwick)		21	41	57	135	185	332	318

NOTES:
*1 The employment noted here is only that directly contracted by Hudsons

*2 The employment noted here is only that directly employed by BP. There are seven base operatives working full time in addition to the only BP employee (the manager) but they have been contracted from another of the supply base companies and are thus accounted for in the figures given (seven operatives have been required to be present on the base since 1974)

*3 The increase in employment at Norscot Services between 1976 and 1977 was caused by the development of the engineering side of the base's services, for the most part. This reflected the changing pattern offshore, from the exploration to the development phase of some of the major oilfields. The numbers of people in the 24 client companies at Norscot are not included here, as they are not base employees

*4 1972 is taken as the base for employment at Lerwick Harbour — before the oil developments really affected the situation greatly. Therefore any additional employment since 1972 is considered to be oil related. The increase in the staff can be gauged by the fact that in 1972 there was only one part-time pilot working at Lerwick Harbour — there are now five full-time pilots. In 1975 a further increase was occasioned by the introduction by the Harbour Trust of a 24 hour port service, which required an additional four port control operators to be employed. Other increases in staff are accounted for by office staff, labourers and boatmen

SUMMARY: It can thus be seen by the above figures how employment which is directly maritime oriented has been increased because of the oil related developments.

From 1978 to 1980 there has not been any significant change in employment numbers in the Marine Services.

Employment at O.I.L. in Lerwick increased greatly in 1978 when the new base opened. Further increases in employment are expected with the continuing development of services in Shetland.

that developments in and around the Harbour have been very considerable since 1970. There have been several stages in these developments and they could perhaps be summarised as follows:—

Stage 1

This was a scheme which involved the Lerwick Harbour Order 1971 being passed through Parliament and given the Royal Assent in July 1971. Included in the scheme was the reclamation of the ground at what used to be No. 1 Herring Station, the erection of a new fish market, and the erection of a 240 foot long spur from the north end of Alexandra Wharf. In February 1972 William Tawse of Aberdeen was contracted to carry out the necessary works at a cost estimated at about £190,000. The land between the Alexandra Wharf and the new No. 1 Station development was also filled in and decked over at an additional estimated cost of £47,000. Whilst this development occurred before the Oil Era was fully under way, and was not strictly oil-related, it proved to be a useful improvement, and the beginning of the excellent development programme which has emerged in respect of the Harbour.

Stage 2

A very significant move in the future development of the Harbour came in 1972 with the purchase of the 1500 acre Gremista Farm in the north Harbour area, by the Lerwick Harbour Trust. This acquisition assured that the effective control of the north Harbour, and most of the development land around the Harbour, remained in the hands of the Lerwick Harbour Trust, and not of possible outside speculators. An industrial estate, part of Norscot Base, and other diverse and not necessarily oil linked developments, have been provided on this land.*

Stage 3

In September 1973, Sir Robert McAlpine & Sons Limited were contracted to build the next phase of developments, being a jetty and reclaimed foreshore, which were to provide facilities for Shell UK Exploration and Production, and BP Petroleum Development Limited, for supply bases.

McAlpines were to build a quay of 60 x 25 metres, below Holmsgarth, to be used by Shell for rig servicing and were also to reclaim about three acres to give a quay face of 155 metres running at right angles from the Shell jetty, for use by BP. Work was scheduled to be completed in early 1974 (it was slightly later than scheduled), and Shell and BP were granted a 15 year lease of the Harbour works by the Harbour Trust. They also leased land from the Lerwick Harbour Trust on the Gremista Estate west of Lerwick Power Station for the storage of pipes, and other oil rig supplies.

Stage 4

The next stage of development was to be the building of a roll-on/roll-off berth, and the reclamation of land for a passenger terminal and warehousing, south from the oil service berths at Holmsgarth. In 1974 this work proceeded at Holmsgarth, with Costains constructing the new ro-ro berth.

Although undertaken at different times much of this construction was seemingly simultaneous, as schedules overlapped. For example, during 1975, whilst the extension of the existing quay facilities was completed at Alexandra Wharf, the construction at Holmsgarth of the new quays for Shell and BP, and the ro-ro facilities continued. During 1976 the quay and other facilities at Holmsgarth for Shell and BP were completed, and substantial progress was made towards the completion of the ro-ro berth at Holmsgarth. During 1977 the ro-ro and other general facilities were completed, but this was not to be the end of the story.

Stage 5

In August 1978, an announcement was made by Lerwick Harbour Trust that William Tawse Limited of Aberdeen (members of the Aberdeen Construction Group) had won a contract to construct additional wharfage, a finger jetty, and another ro-ro ramp, in order to provide additional facilities for general, commercial and oil related traffic. When the work is completed, the Harbour will have an enclosed dock area of more than 2.5 acres, which amongst other things, will provide safe berthage for 16 large fishing vessels. The final cost of this project is estimated at £3.3 million.

See Appendix for listing of Gremista Industrial Site clients.

Stage 6

On 6th July 1979, a contract was let to William Tawse Limited at an estimated final cost of £900,000 for the reclamation of 2.5 acres and provision of 170 metres length of quay on the shore side of the new dock created by the Stage 5 development. This work (Stage 6) was completed in December, 1980, making the total berthing in the Harbour (inclusive of Oil Supply Bases) up to 10,000 feet.

A small spin-off from these harbour developments has been the building of a breakwater with the object of providing a small marina to accommodate 120 small boats at floating moorings. This breakwater, constructed with surplus rock from roadworks at Gulberwick, has now been completed with a certain amount of reclamation for parking and an access road. It is expected that the marina development itself will take place over a number of years, as funds become available.

Further Developments Around Lerwick Harbour

The land owned by the Harbour Trust is also being developed further. Two developments of note in this respect have been undertaken by Lerwick Harbour Trust:—

(i) **Casing Yard for BP Petroleum Development Limited**

In July 1980, a contract, valued at approximately £340,000, was let to Sullom Quarries Limited to reclaim, by January 1981, two acres of land. This is in order to provide a pipe and storage yard for BP Petroleum Development Limited who, as stated, are expanding their supply base activities at Lerwick.

(ii) **Black Hill Industrial Estate Limited**

This is a joint venture between BP and Lerwick Harbour Trust for the provision of ten serviced sites in the form of a small Industrial Estate. The contract was let to William Tawse Limited in July 1980, for completion in December 1980.

Developments have also been undertaken by organisations acquiring or leasing land from Lerwick Harbour Trust. The major development in this category has been the acquisition of 10.79 acres of the Gremista Estate in 1976, and a further 2.68 acres in 1980, by the Shetland Islands Council, to form the Gremista Industrial Site (see Appendix to Chapter Three). Only one company on this site is involved in work which is wholly dependent on offshore oil activities — Schlumberger Inland Services Inc.

Schlumberger Inland Services Inc.

Schlumberger Inland Services Inc., is an international company quoted on both the Paris and New York Stock Exchanges, whose primary business is oil well logging. The company was first present in Shetland on the Norscot Base in 1978. The pace of developments offshore caused the company's operations in Shetland to expand and in September, 1979 they moved to a new purpose built workshop on the Gremista Industrial Site. This has been one of the most important oil related developments at Lerwick Harbour, and represents an investment of £7 million — £1 million being spent on the workshop, £6 million on equipment.

The company as a whole forms a multi-national group specialising in the measurement and recording of all types of data. It employs over 50,000 people worldwide who produce control systems and electronic equipment, as well as work in the petroleum industry.

The principle function of the company in Shetland is to provide electrical logging or wireline services. Wireline services are as vital to the exploration for and production of oil and gas as the x-ray is to medical diagnosis.

To find the oil and gas reservoirs it is first necessary to locate geological areas where they might be present. Seismic testing (shock waves) give preliminary indications of structures which may contain oil but there is no reliable way to measure precisely these subsurface structures without drilling.

Even when a well is drilling, very little information is available to the geologist standing at the top of a hole which is several thousand feet deep and only a few inches in diameter. Often there is no evidence at the surface that the well has penetrated an oil or gas reservoir.

To obtain information about the structure, drilling is interrupted periodically to allow a Schlumberger mobile laboratory to lower various measuring instruments to the bottom of the drill hole on a line of wire-wrapped electrical cable — hence the term "wireline". As each instrument is pulled out of the hole it measures, using electricity, sound waves and radiation, the depth and physical properties of the various formations it passes. These measurements are transmitted on the wireline to be recorded in the mobile laboratory. The recorder in turn produces a graph called a "log", with various combinations of logs revealing a complete picture of subsurface formations to show how deep, how thick and how porous the rock is, and its gas and oil content. The information obtained from logs is thus essential to:—

(a) the locating of reserves of oil and gas;

(b) evaluating the production potential of a well once an oil or gas bearing formation has been located;

(c) deciding the location of future wells.

The working team for wirelining services consists of an engineer and several operators. The engineer will have a degree in engineering, electronics or possibly goelogy, and will have passed through specialised Schlumberger training schools. The operators act as back-up to the engineers and are skilled members of the team, it taking a period of three years on-the-job training for operators to acquire complete competence. The operators must be able to maintain the range of Schlumberger tools, rig up the tools for a down hole survey, operate the winch control during the survey, respond to tool failures, mechanical breakdown and other emergencies, and generally assist the engineer in running the log and developing film for the client.

Logging is a service job undertaken for the oil companies and is organised on an on-call basis to proceed so that expensive lost rig time is minimised. Once started, the Schlumberger team continues working until the job is finished — which can be up to three days, in extreme cases.

The work cycle therefore depends on the wells being drilled offshore, and because of this, is not the regular two weeks on, two weeks off pattern of other offshore work. On average, Schlumberger employees can expect to spend 14 to 18 days of every month offshore. Compensatory time is given for time worked offshore. The company presently (January 1981) employs 92 people in Shetland, 24 internationally and 68 locally, of whom six are female.*

Shipping Services based at Lerwick Harbour which are affected by Oil Related Developments

As can be imagined, all shipping services based at Lerwick harbour have benefitted from oil related developments.

Shipping services include owning ships, chartering ships, and acting as agents for other shipowners. There are three shipowning companies operating regularly to and from Lerwick Harbour:—

1. Hay and Company Limited;
2. P&O Ferries Limited;
3. Shetland Line Limited.

1. Hay & Company Limited

Hay & Company Limited are one of the oldest trading companies in Shetland with their origins being traceable to the end of the eighteenth century. The company have been shipowners and fishing boat owners since 1840. Their activities at Freefield, where they built and own their own dock, sawmill, coalyard, builders merchants and ancillary warehousing, have gradually developed through the years.

After a lapse of about 40 years, Hay & Company resumed shipowning in 1974 and own two vessels named "Lerwick Trader" and "Shetland Trader". The present Shetland Trader (bought in 1979) is the flagship, at 1133 tonnes deadweight, the "Lerwick Trader" being 888 tonnes.

The advent of oil has brought many changes to this company and its operations. Since 1972 the Freefield dock area has been transformed, with new buildings being erected to cope with the growing demand for timber and other building materials.** In respect of the shipping activities in 1978 their smaller 600 tonne cargo vessel was sold and replaced by a vessel that loads over 1000 tonne cargoes. More goods and cargoes have been imported to

---

*All the women are employed onshore.
**See Chart XI "Imports of Cement, Coal and Timber to Lerwick Harbour".

Shetland by the ships, and they trade less between ports elsewhere as they did formally. Offices have been opened at Sullom Voe and Sumburgh Airport by Hay & Company, to provide agency services to both shipping and the air services.

In 1975 another change occurred when the company was acquired by the timber and builders merchants, John Fleming & Company of Aberdeen.* All the timber required for the construction of the Sullom Voe Terminal has come through Hay & Company. The two "Traders" bringing shiploads of timber directly from Finland, Sweden and Norway. Other ships are also used to import timber from the Baltic.

IMPORTS OF CEMENT, COAL AND TIMBER TO LERWICK HARBOUR*[1] Chart XI

Year	Cement*[2]	Coal	Timber
1972	7079	4835	2358
1975	2770	4404	2095
1976	14393	4809	1858
1977	13505	4296	3427
1978	19458	4562	2883
1979	16633	6103*[3]	3184
1980	11466	5892	1834

*1 All these imports were made by Hay & Company Limited. The coal was imported in their own ships; the cement and timber partially by their own ships, partially by other ships (see text).

*2 Cement is the commodity showing the greatest rise from the baseline year of 1972, before onshore developments began. The figures reflect the major construction activities at Sullom Voe and Sumburgh.

*3 1979 was a poor year for peat cutting due to inclement weather, partly explaining rise in coal figures.

Hays are sole importers of Blue Circle cement. Bagged cement is imported in their own ships and bulk cement in other ships which are sent directly into Sullom Voe. Coal imports to Lerwick Harbour have also increased a little with the rise in population and prosperity that has occurred, but can still fluctuate a little with the weather — if it is a good year for peat cutting, less coal is used.

Hays activities have also greatly increased, with the rise in the number of vessels calling at Lerwick Harbour and the opening of Sullom Voe as a port. The company have expanded their activity as shipping agents, to handling goods, people and customs needs for the oil industry at Sumburgh and Unst airports.

A measure of the expansion of Hays activities is that in 1972 they had 104 staff; in 1980, 150, of whom 23 were women. All are locally employed in Shetland.

2. P&O Ferries Limited

In 1971 the North of Scotland, Orkney & Shetland Shipping Co. Ltd. became part of the P&O Steam Navigation Company Group when that Company bought over Coast Lines Limited. The Company's name was then changed to P&O Ferries in 1976.

---

*Flemings also own the Sumburgh Hotel and are part owners of Hagdale Lodge Hotel in Unst.

Since 1971 there have been very great increases in both the number of passengers and the amount of cargo carried by P&O's ships. For example:—

PASSENGER JOURNEYS ABERDEEN/LERWICK/ABERDEEN

Year	1970	1975	1979
No. of Passengers	22092	34875	52821

For 1972 to 1979 the tonnages of cargo carried have been estimated to have increased at least threefold, but an exact measure is hard to estimate as the mode of operation of the ships has changed from lift on, lift off to roll-on, roll-off (ro-ro).

Under the old lift on, lift off system the cargo was groupage — i.e. loose cargo was delivered to P&O's Aberdeen depot, checked and stowed in the Transit Sheds, and thereafter mainly barrowed or carried by fork lift truck to the vessel's side, from where it was slung on board by ropes and wires in conventional fashion.

With the ro-ro system most cargo is driven straight on board the ship — in vehicles of varying capacity, some carrying 20 tonnes. There is still groupage cargo, and it perhaps gives the best reflection of trends. Such cargo is loaded on board P&O's Slave Trailer equipment and the important Exports of Livestock carried in the GLT's (Goods/Livestock Trailers) designed by the Company.

For example:

GROUPAGE CARGO

Year	1972	1979
Tonnage (Metric tonnes)	41251	37095

Groupage cargo was down in the first half of 1980 by 10.36% reflecting perhaps the fact that construction at Sullom Voe was slackening as the Terminal became nearer to completion.

The change from lift on - lift off operations to roll on roll off required the replacement of the ships used to serve Shetland.

The 'new'* ships were also bigger in terms of deadweight size than the old. A basic comparison between the ships used is seen in the following chart:—

CHANGES IN P&O FERRIES SHETLAND FLEET

Ships (old)*1	Gross Registered Tonnes	Passengers Carried	Ship (new)*2	Gross Registered Tonnes	Passengers Carried
St Clair	3303	500 summer 350 winter	St Clair*3	4468	730
St Clement	815	12	St Magnus	1205	12
St Rognvald	941	12	Rof Beaver	983	None

NOTES:

*1 'Old' means before ro-ro i.e. pre 1977.

*2 'New' means since ro-ro established in 1977.

*3 A major difference since 1977 is that the main passenger ship, the St Clair, now runs three times per week (since October 1979) instead of two times per week previously. The 'new' St Clair began operations in April 1977.

*The ships were 'new' to Shetland although not newly built.

Ro-ro operations were made possible by the building of a new berth at Lerwick and began on the 18th June 1976, when the Rof Beaver first used the facility, although the new St Clair was not available until April 1977. P&O's new Ferry Terminal was not officially opened until 15th April 1977, when the then Secretary of State for Scotland, Bruce Millan, did the honours.

A further expansion that has occurred for P&O Ferries is in the size of their Shetland haulage fleet which has increased from eight to 24 units, with the number of drivers increasing from eight in 1972 to 20 in 1979. A garage was opened at the North Ness, specifically to service these vehicles and the first mechanic was employed in 1974 (there are now three full-time mechanics and one apprentice there). The increase in cargo has also led to an increase in the number of checkers employed, from eight in 1972 to 12 in 1980. The construction of the Sullom Voe Terminal also engendered extra business.

Special sailings to the Construction jetty at Calback Ness by the cargo ships began in early 1974, although it was not until September 1977 that P&O established a separate office at Sullom Voe.* Up till then, the Lerwick office had sent staff to Sullom Voe as sailings dictated. From May 1975 until the end of 1979 approximately 70,000 tonnes of cargo were shipped directly into Sullom Voe. The peak years were 1976 and 1977, averaging about 20,000 tonnes per year; in 1978, 18,927 tonnes were shipped and in 1979, 14,653 tonnes.

By 1980 P&O Ferries employed 58 people in their Shetland operations, the majority based at and involved in the life of Lerwick Harbour. Since the oil-related activities had begun P&O have moved (in 1977) from their old warehouse at Victoria Pier to the new Holmsgarth Terminal, where they have a transit shed of 1,248 square metres and 252 square metres of office space — a considerable change from what there was before!**

A new development under the P&O flag is the Oil Tanker Agency Service now offered at Sullom Voe at which location two members of staff are engaged on a full-time basis.

3. Shetland Line

Shetland Line Limited, one of 27 members of the Melton Securities Group, began a shipping service to Shetland as oil developments materialised and the demand for imports was stimulated.

The company started in a small way with Norscot Services Limited in Lerwick acting as their agents. In early 1975, as business expanded, Shetland Line established its own office personnel at Norscot Base, rented warehousing and utilised base operatives for cargo handling. Continuing expansion led Shetland Line to move out of Norscot to Magray Yard and warehouse at Gremista in 1976.

About the same time (1976) Shetland Line started to use J. & M. Shearers Quay at Garthspool, Lerwick, and placed a crane there to aid operations. Containerised cargoes were moved between Shearers Quay and the Gremista warehouse and yard until finally, in September 1978, the company moved and rented warehousing from Shearers at Garthspool.

At Garthspool, Shetland Line lease approximately 5000 square feet of warehousing in addition to open storage and offices, adjacent to the quay, where the company has priority on the berth. Shearers do all the stevedoring for Shetland Line.

Shetland Line operate shipping services in three capacities (as do Hay & Co. and P&O Ferries) namely:—

(i) as shipowners;

(ii) as charterers of ships;

(iii) as ships agents.

(i) **As Shipowners**

The biggest single development of the company to date was in buying ships. In March 1979 the Melton Pioneer was acquired to ship timber from Sweden to the UK; thereafter the timber and container ship

---

This office was in a temporary portacabin, building to cover the construction phase at Sullom Voe with Wimpey Marine Limited provided the agency services for P&O's ships at Sullom Voe construction jetty.

**For example there is now a passenger waiting facility with a snack bar and other accommodation whereas there were no snack facilities before.*

Melton Challenger, newly built in West Germany which mainly runs from Canada to the UK. The Melton Clipper engages in general North Sea traffic, but it is the appropriately named Melton Viking of 1300 deadweight tonnes which serves Shetland, providing a weekly service from Grangemouth.

The Company has three groupage cargo depots — one at Bellshill (Scotland), one at Grangemouth docks and one outside the port of Grangemouth. All containers are emptied and goods delivered in Lerwick by the company's own personnel and haulage.

(ii) **As Charterers of Ships**

The chartering of ships to carry cargo to Shetland was perhaps Shetland Line's first noticeable activity in the Islands. One of the most interesting examples of this activity occurred in 1974 when Shetland Line chartered the Navipesa Dos, a Spanish owned ro-ro ship, to go into Calback Ness with JMJ's earth moving equipment for construction work at Sullom Voe. The Navipesa Dos backed into the beach at Calback Ness (the ship was carrying equipment to help build the Construction jetty), two diggers rapidly drove off, and then the ship was roped to the two diggers in order to berth, and discharged the rest of her cargo.

Not all oil related ship charters are quite as exciting as that but Shetland Line is known for chartering specialised ships for special tasks. Other examples of activity include the shipping up to Shetland of the construction camps at Firth and Toft. The Firth camp came through Lerwick Harbour in packages; Toft camp was shipped directly to Toft from Boston, Lincs. Shetland Line played a major part in shipping construction materials such as steelwork, for the oil related airport developments at Sumburgh Airport, developments which included the new £16 million terminal completed in 1979.

Shetland Line also chartered a specialist ship which had no encumbrances in her hold, to ship in the new extension to the Lerwick Hotel in 12 ton units. Everything except the beds, colour televisions and telephones were in the units (the beds came in containers). This happened on 21st July 1980; by September the units were in use.

All chartering for Shetland Line is done from Melton Mowbray. In Shetland it is shipping operations which are concentrated upon, the main base being at Lerwick Harbour. One operations supervisor is permanently at Sullom Voe to handle all construction traffic but the ships are operated and discharged by Wimpey Marine with the Shetland Line supervisor there simply to take care of clients interests, and co-ordinate distribution of cargo.

(iii) **As Ships Agents**

Shetland Line act as agents for their own ships and for ships they have under charter. They also do pure ship agency work for small cargo vessels* mainly for German shipowners with business in Shetland.

Shetland Line had a total of 12 employees in Shetland at the end of 1980, all but one being local people. It is the one Shipping company that came to Shetland with the Oil Era and has stayed. Others, such as Ellerman Wilson, Olsen, Everard, Orkney and Shetland Carriers, and Fife Traders, have come for a short while and gone again, suggesting that there is room for the two groupage and general cargo services** now operating from Lerwick Harbour but perhaps not for more.

General Ships Agency Services

Several ships agency services are available also at Lerwick Harbour through Norscot Services, and Ocean Inchcape (Shetland) Limited — they act as agents for their own Ocean Inchcape Supply Boats — and J. and M. Shearer Limited.

Ships agency work for tankers is specialised.
**i.e. Shetland Line and P&O Ferries.*

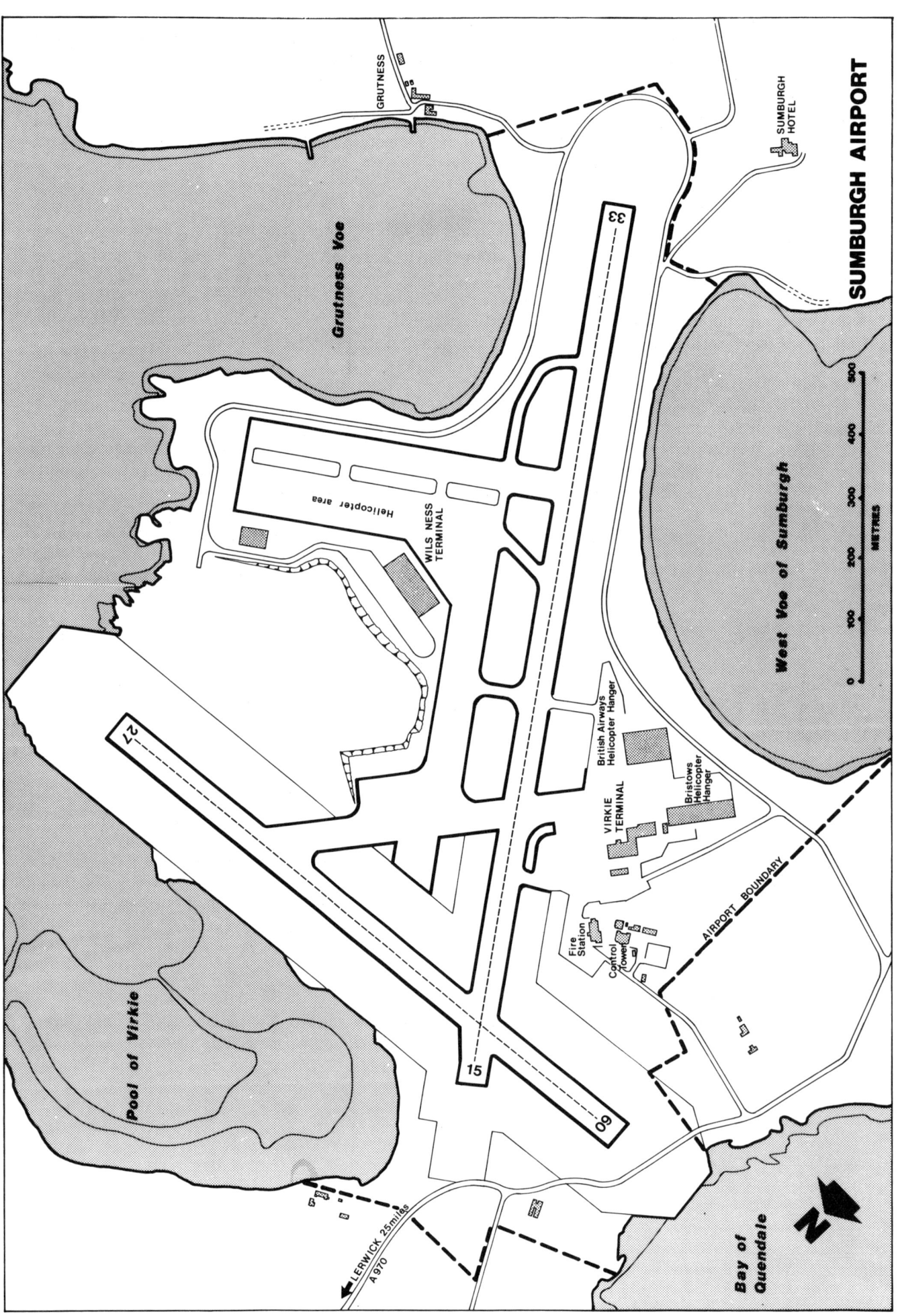

Chapter Four
Air Services

A tremendous growth in air traffic has occurred in and around Shetland, since the search for oil and gas began offshore in 1971. In Shetland, this traffic was concentrated almost exclusively at Sumburgh Airport, situated at the most southern part of the Mainland (see map), from 1971 to 1978. Developments began to spread further north to other parts of Shetland in 1978, so that by 1981 there were four oil affected airports in Shetland — at Sumburgh, Tingwall, Scatsta and Unst — each playing a different role in the pattern of activity.*

The development of air services could perhaps be considered to fall into two categories:

I the development of the oil related airports in Shetland;

II the development and integration of flight advisory and control services to cope with the air traffic generated both by the airports on Shetland and by the platforms, rigs and other oil related vessels offshore Shetland.

I. **The Development of the Oil Related Airports in Shetland.**

  (a) **SUMBURGH AIRPORT**

  (i) **Background**

  Flat land is scarce in Shetland, hence it is not surprising that a major airport has developed at Sumburgh, which has one of the few stretches of flat land in the Islands. Sumburgh was established as an airport long before oil developments began. Scheduled passenger services were introduced by Allied Airways on 2nd June 1936, operating Rapide aircraft from a grass strip. A rival company, Highland Airways, commenced operations on 3rd June 1936, and until the outbreak of the Second World War in 1939 there were three return flights a day linking Sumburgh to Orkney, Wick, Aberdeen and Glasgow. Throughout the Second World War from 1939-45, Sumburgh had a daily service link to Aberdeen which was operated by unarmed aircraft and in radio silence.

  The Royal Air Force (RAF) established a Coastal Command and communications base at Sumburgh at the outbreak of the war. The airport was then exceedingly busy at times, with squadrons of Mosquito fighters and Lancaster bombers as frequent visitors. Sorties were made by aircraft from Sumburgh to Norway and on protection duty for the convoys to Murmansk. The RAF constructed three concrete runways, two of which are still in use, and a variety of outbuildings, including a control tower, which continued to be used long after the war ended.

  In 1947 British European Airways (BEA) took over the scheduled services, operating a Dakota DC-3 aircraft until 1962 when Dart Heralds were introduced. In 1966 runway 15/33 was extended to permit the use of the larger Viscount aircraft for the scheduled services, and an improved terminal building was constructed.

  During these years after the Second World War and before the Oil Era, life at Sumburgh was peaceful. Ponies grazed on the airport and a golf course was eventually established for a short time — games of golf being temporarily halted by a red light on the Control Tower when a aircraft was due.

See Chart XII "Oil Related Airfields in Shetland"

OIL RELATED AIRFIELDS IN SHETLAND   Chart XII

Name	Location & Elevation	Ownership	Runways	Runway Length, Direction and Type	Navigational Aids
SUMBURGH	*Location:* Lat: 59° 52' 45" N; Long: 01° 17.50' W 17 miles (by air) South of Lerwick. *Elevation:* 18 ft. above mean sea level	Civil Aviation Authority	3	1. *Runway 09/27* 1084 x 46 metres 083°(T) - 263°(T) 091°(M) - 271°(M) Tarmac 2. *Runway 15/33* 1426 x46 metres 143°(T) - 323°(T) 151°(M) - 331°(M) Tarmac. 3. *Runway 06/24* 550 x45 metres 053°(T) - 233°(T) 061°(M) - 241°(M) Tarmac. 06/24 is for helicopter use only	1. Non Directional Beacon (NDB) 2. Doppler VHF Onmi directional Range (DVOR) and co-located Distance Measuring Equipment (DME). 3. Plessey 430 Radar — for surveillance radar approaches to ½ nautical mile on Runways 09 and 27. 4. Marconi 4032-23 cm primary radar for long range surveillance to 120 nautical miles. 5. Cossor 9000 Secondary Surveillance Radar (SSR) for long range surveillance to 180 + nautical miles. 6. Radio: Tower 118.25 Mhz; Approach: 125.15 Mhz, 119.25 Mhz; Radar: 130.05 Mhz; Automatic Terminal Information Service (ATIS) 125.85 Mhz; Emergency 121.5 Mhz.
TINGWALL	*Location:* Lat. 60° 12' N; Long. 1° 14' W 4½ miles north west of Lerwick. *Elevation:* 13.7 metres above mean sea level.	Shetland Islands Council	1	*Runway 20/02* 740 metres; 200°(M)-020°(M); Tarmac.	1. Non Directional Beacon (NDB) Radio; 122.6 Mhz
SCATSTA	*Location:* Lat. 60° 26' 3" N; Long. 10° 18' W 21 miles N of Lerwick.	Shetland Islands Council and Frank Hunter are land owners; Airwork Ltd. are managers for BP.	1	*Runway 06/24* 960 metres; 060°(M) - 240°(M) Tarmac.	1. Non Directional Beacon (NDB). 2. Plessey 430 Radar for surveillance approaches to height of 430 feet within ½ mile of the runway.
UNST Baltasound	*Location:* Lat. 60° 45' N Long. 00° 51' W 41 miles NE of Lerwick. One nautical mile south east of Baltasound.	Shetland Islands Council — operated by Airwork Ltd.	1	*Runway 12/30* 610 metres; 120°(M) - 300°(M) Tarmac.	1. Non Directional Beacon (NDB). 2. Radio Frequency — 130.35 Mhz.

ire & Rescue	Lighting	Other Facilities	Hours of Opening	Comments
tenders — 650 gals water and foam; 72 kilos C02; Inshore rescue raft; ambulance	*Runway 09/27* — High intensity runway and approach lights. Visual Approach Slope Indicators (VAI) and Precision Approach Path Indicators (PAPIs) *Runway 15/33* — High intensity runway lights. HI strobe on approach to Runway 24. *Taxiway* — green centre line. *Aprons* — Flood-lights. *Obstruction Lights* — Red.	3 windsocks, 5 helicopter hangars, fuelling; de-icing; 1 fixed wing hangar; stand-by generators; 2 terminal buildings — baggage handling facilities; cafeterias, shops, bank, car hire, public telephone, booking and handling agents, check-in facilities. Tel. Sumburgh 60654	Mon-Fri 0715-2200 Sat. 0800-1800 Sun. 0900-1700 and on request. Opening minimum 3 hours. £450 per time (1980 money).	1975 — £7 million on improvements. 1977 - £5 million of further improvements. 1978/79 — £16 million on new Terminal, helicopter runway and other facilities.
fire tender;	Battery lighting Litas approach units (low cost visual approach aids); Apron lighting; Parking area lighting.	1 windsock; fuelling bowsers — Jet A1 and other fuels; offices, public telephone, waiting room and toilets. Tel. Gott 306	Mon-Fri 0900-1700 and as required. Prior permission needed for all aircraft.	This airfield was built in 1976 at a cost of £250,000, replacing a grass strip.
fire tender water tender; ambulance services from Sullom Voe Terminal.	Approach lighting; obstruction lighting; beacons as required by Civil Aviation Authority; aerodrome beacon, Precision Approach Path Indicators (PAPIs)	1 windsock; control tower and equipment rooms; workshops and stores; stand-by generator and switch gear building; Snow-blower; Konsin de-icing spreading machine; 2 snow-ploughs — large and small; Konsin bulk storage tank farm; capacity 13000 gals (450 gals used in each run down runway); waiting room; carparking; toilets; offices. Tel. Sullom Voe 2487.	As required, prior permission only. Mon-Fri 0800-1800 normally.	£2 million spent on improvements in 1978 in order to bring airfield up to standard to take HS748 aircraft; first aircraft landed 25th July, 1978, (fog at Sumburgh — diversion).
fire tender Landrover 00 gal. water/ oam. First Aid Centre.	High intensity airfield lighting, approach guidance by Precision Approach Path Indicators (PAPIs)	1 windsock; Terminal building, refuelling for fixed wing, and helicopters, helicopter hangar; offices; toilets; public telephone Telelephone: Baltasound 404/5/6/7	Mon-Fri 0830-1630 or as required by prior permission only.	This strip was improved by an oil company consortium managed by Chevron in 1978/79 to meet C.A.A. requirements for Twin Otter aircraft and to provide facilities for fixed wing to helicopter crew change at a cost of approximately £1 million. A helicopter hangar was completed for Bristows in June 1980 at a cost of £1 million.

(ii) **Air Traffic**

One of the first indications of the changes to come was seen in a significant increase in air traffic movements from 1970 to 1971*. This increase reflected the beginning of the exploration for and the discovery of the oil and gas resources offshore Shetland. From 1970, until 1978, when there was a peak in activity at Sumburgh, both aircraft and passenger movements increased more than 20 fold. There was also a very dramatic increase in the amount of freight handled.**

(iii) **Improvement of Facilities**

In 1970, a new terminal building had been constructed, and this was extended in 1972 to try and cope with the greatly increased air traffic. When the Civil Aviation Authority (CAA) took over the Airport from the Department of the Environment in 1973, these facilities were proving quite inadequate to handle the emerging situation. Even as late as 1974, scheduled passenger services had to be restricted to the hours of daylight because of a lack of navigational aids, including lighting. A decision was thus taken in 1974 to develop one of the two runways then in use — 09/27 — to allow instrument landings to be made by aircraft. Sophisticated navigational aids, runway lighting, and generally improved facilities, including a further extension to the airport terminal, were planned. These improvements were built, and completed in 1975, at a cost of approximately £7 million (1975 money). Hardly had these improvements been completed, when, it was announced that the CAA, as managers of the airport, were going to undertake further improvements in the form of additional fire service facilities, meteorological offices (a full weather forecasting service was to be installed),*** aircraft parking space, etc. — to a total cost of £5 million. Work on these facilities began in the Spring of 1977 and was completed by the end of the year.

Meanwhile, both Bristows and British Airways Helicopters had been expanding their hangar areas, it being estimated that each of these helicopter companies spent in excess of £2 million on new hangars and other facilities for servicing aircraft at Sumburgh Airport. By 1978, therefore, an investment of approximately £16 million had been made at Sumburgh Airport, in new facilities.

This was by no means the end of the expansion. Projections of the increased traffic flow at Sumburgh had been made which suggested that up to 700,000 people might pass through the airport in 1978, the majority being oil related helicopter/fixed wing transit passengers. The Civil Aviation Authority in 1977, decided to conduct a "Sumburgh Airport Passenger Survey",**** to analyse what was becoming a situation of intolerable overcrowding of people, and an overload on the available services. This survey produced several interesting results, the most notable being that 73.1% of all the people using Sumburgh were oil related helicopter crew change workers. Sumburgh had become a staging post for helicopter to plane, and plane to helicopter movements, the report noted. Of the remaining people using the airport at that time, 14.1% were using Sumburgh for access to Sullom Voe, and 12.6% were 'other' terminating passengers. Looking more closely at the 26.8% or so of the airport users — the passengers terminating in Shetland — the Survey found that 53% of all terminating passengers were Sullom Voe workers, 35% were people bound for Lerwick, and the remaining 12% were bound for outlying villages. Thus the main areas of both growth and congestion were to be found in the oil-related sector of offshore workers, and the Sullom Voe workforce (in a lesser proportion).

At a meeting of the Sumburgh Airport Consultative Committee (a Committee which had been formed to tackle the problems being experienced at the airport, and consisting of representatives from the Shetland Islands Council, Dunrossness Community Council, the CAA, the helicopter operators and the main fixed wing operators at Sumburgh),***** in January 1978, it was indicated to the Shetland Islands Council and other interested parties, by the CAA, that development proposals for the airport, outlined in November 1977, had been totally revised. Detailed investigation had revealed operational problems, and forecasts had indicated that oil related traffic would increase at a high rate. A plan was therefore being considered which would channel the

---

Helicopters started flying out of Sumburgh in 1971, see later in chapter.
**See Chart XIII "Sumburgh Airport Usage"*
***Meteorological services began at Sumburgh on 1st March, 1978, with a 24 hour coverage, employing seven people on shift work, being provided. By 1981 these services had expanded further, to employ ten people full time.*
****See 'CAA Sumburgh Airport Passenger Survey, 1977', published by the CAA, London, in June 1978.*
*****See Appendix for full details of this Committee's representation and functions.*

SUMBURGH AIRPORT USAGE

Chart XIII

Year	Total Aircraft Movements*1	Total Passengers*2	Total Freight (kg)	Breakdown of Aircraft Movements Fixed Wing Scheduled	Fixed Wing Charter	Helicopter	Other*4
1966	1238	23856	N/A*3	N/A	N/A	N/A	N/A
1969	1896	26017	N/A	N/A	N/A	N/A	N/A
1970	2386	32068	N/A	N/A	N/A	N/A	N/A
1971	3895	46944	N/A	N/A	N/A	N/A	N/A
1972	6095	70589	205303	N/A	N/A	N/A	N/A
1973	9114	94312	570908	2537	3290	3173	114
1974	16966	158253	931327	3265	5373	7273	1055
1975	22224	186319	1101007	3069	7698	9577	1880
1976	25694	250778	1574338	3086	9455	12664	489
1977	38753	404370	1815392	3725	15441	18496	2342
1978	50666	685492	1950287	3719	20124	26823	319
1979	47771*5	622446	3274754	4116	17621	26034	694
1980	41307	608169	3801467	3946	13656	23705	200

Notes:
* *1 An aircraft movement is defined as either an aircraft landing or an aircraft taking off, e.g. one aircraft landing and taking off again = two movements
* *2 Many of these passengers are in transit as oil rig crew changes.
* *3 N/A = Not available. Detailed figures often only available since 1973 (or later) when Civil Aviation Authority took over management of the airport.
* *4 'Other' includes military and private planes — such as flying club flights.
* *5 Decrease in figures from 1978 reflected change in pattern of activity with the development of Tingwall, Scatsta, and Unst airports. The number of movements have declined more than the number of passengers pro rata reflecting better utilisation of aircraft.

oil related helicopter traffic to the Wils Ness area of the Airport, by the building of a new heliport. Improvements to the existing mode of operation were detailed as follows:

(a) the road crossing the runway 15/33 would be removed, following the construction of a new road located entirely outwith the airfield; a new extension from the road would connect with the proposed new terminal;

(b) the existing Wils Ness apron, brought into use in 1977, would be expanded in a south-easterly direction, and would enable additional fixed wing aircraft to be parked;

(c) a new passenger terminal would be built, which would serve as an oil related passenger transfer building from fixed wing aircraft to helicopter, and vice versa;

(d) helicopter parking would be available east of the new passenger terminal and a link taxiway would connect to a new helicopter runway south of the new passenger terminal abutting runway 15/33, almost at right angles. (See Diagram). The CAA felt that Sumburgh Airport would reach the finite limit of development with these proposals, with the resulting facilities being completed by the summer of 1979.

After the introduction of these proposals at the Airport Consultative Committee in January 1978, steps were taken to put them into practice. In February 1978, planning permission was applied for, and in April 1978, the CAA were given planning consents by the Shetland Islands Council, once worries about the development at Unst Airport had been assuaged.

In the following month, May 1978, the first of the contractors, Costarc* Joint Venture Ltd., arrived on site, followed by G. Percy Trentham at the end of July. By 13th November 1978, the apron, runway and road works were completed, and by December the airfield ground lighting works were completed. The Terminal building itself was completed by July 1979 — little more than a year from the commencement of the construction project — and officially opened by Her Royal Highness Princess Alexandra on 28th September 1979. The building of these facilities within such a short time was a major achievement, especially given that the site had to be blasted out of the rock face before construction could commence, and that meanwhile full scale flying operations continued at the airport. The new Wils Ness heliport, with parking for up to 20 Sikorsky 61 N helicopters and for three Viscounts or Hawker Siddeley 748 types of aircraft, eased the over pressured situation in the old Terminal building, which reverted to its former use for fixed wing scheduled traffic. The new building, some 6500 square metres in area, was designed specifically to cater for the all male oil related traffic. Due to the over crowded situation in the old building the transit of passengers from one aircraft to another had been taking up to three hours — the new building transformed changeover times to less than one hour. The heliport is also perhaps unique amongst airports, in that the majority of passengers arrive and depart by air.

The building itself was built using the Conder prefabricated Kingsworthy system of building, where a series of pre-designed components — steel columns, concrete wall panels, windows and metal roof decking — was arranged to suit the architect's layout. This allowed a dry 'envelope' or shell building to be rapidly erected and provided a reasonably protected enclosure for the other trades to work in during the Shetland winter. There is triple glazing, and high relief murals form acoustic panels in the main concourse, to combat noise. The building is designed to accommodate a peak loading of 800 men plus 150 office staff at a time, and has closed circuit television, flight information, and public address systems to aid the efficient transit of passengers. An all electric 250 meal kitchen is provided on the first floor, where Grand Metropolitan Airport Services run a cafeteria. There is also a well stocked shop and other facilities in the Terminal but no bars, as alcohol in totally banned in the interests of offshore safety. The building also houses some 30 companies in modern office accommodation.

Other Improvements at Sumburgh

(a) **Emergency Services**

Part of the Civil Aviation Authority's improvements programme at Sumburgh included a new four bay fire station, combined with a two bay motor transport workshop which was completed on 30th October 1978, at a cost of £350,000. A further £150,000 has been spent on two new Foam/CO_2 fire tenders. The number of firemen at Sumburgh has also expanded from six to 25 people to provide full cover daily from the hours of 0630 to 2230. A new 25,000 gallon water storage tank, automatic booster pump and extended fire main, and automatic fire alarm system, were required to be installed to comply with fire safety regulations in the new passenger building, aircraft parking and fuelling areas.

A joint venture company formed by Costains and the Amey Roadstone Company.

(b) **Telecommunications**

In line with the physical development of Sumburgh Airport an extensive programme of renewing and updating telecommunications equipment has been undertaken.

When the extension and upgrading of runway 09/27 occurred in 1975 a Plessey ACR430 primary radar was installed to support the existing Distance Measuring Equipment, Doppler VHF Omni-Range and Non-Directional Beacon.* This radar became operational early in 1976.

In 1978 an Automatic Terminal Information Service came into operation. The service continually broadcasts the essential information — visibility, wind speed and direction, air pressure, and runway conditions — needed by a pilot, as his aircraft approaches. Previously this information had to be given verbally over the radiotelephone by an air traffic controller. This information is now tape recorded every half-hour by a controller or his assistant, who have the latest meteorological reports displayed on closed circuit television. The incoming pilot receives this information by tuning to the correct radio frequency.

Automatic Direction Finding (ADF) equipment has also been installed in the Control Tower to give an aircraft's bearing in a digital read-out as soon as the pilot has contacted the Control Tower.

To enable radar approaches from the west as well as the east, a second electronic marker was installed to increase the efficiency of the ACR 430 radar, which was relocated to make way for further developments.

Other works included a new Visual Control Room (VCR) and Approach Control Room together with new radar equipment. The Visual Control Room and a service module to contain the telecommunications equipment room and air conditioning plant were mounted on a short stalk tower adjacent to the old Control Tower. The VCR formed part of a £1.2 million contract awarded to G. Percy Trentham for work in connection with the improvement of the Control Tower. In addition to the VCR, a new Approach Control Room was built in front of the old Control Tower with access by covered walkways. The Control Tower itself now houses National Air Traffic Services (NATS) administration and telecommunications facilities.

New radar systems include a primary radar supplied and installed by Marconi Radar Systems and a Secondary Surveillance Radar** installed at Fitful Head by Cossor Electronics.

The Cost of the Developments at Sumburgh

It is estimated that since 1974 more than £32 million has been spent developing Sumburgh Airport. To help finance these developments, the Civil Aviation Authority was assisted by a grant of up to £2.4 million from the European Regional Development Fund, and by a loan of £10.8 million from the European Investment Bank. The cost of the development is being recouped over a ten year period by increased airport charges. This has acted as a considerable disincentive to some airport users and has stimulated the use of the other oil related airports in Shetland where costs are lower. Sumburgh has not the passenger throughput of Aberdeen to keep costs per head or per aircraft lower, yet because of its location and the nature of its operations, requires the navigational aids and other facilities normally only pertaining to major and commercially viable airports.

The cost of operating Sumburgh is currently in the region of £5 million per annum, and it is to be hoped that a solution is found to the dilemma in which the airport management has been placed in respect of being obliged to levy high landing charges. This Airport plays a very vital role, not only in the services it offers to the aircraft and people passing through, but also in the Air Traffic Control, advisory and navigational services it provides for the very extensive traffic now operating in the Shetland area.

See Table IX "Navigational Aids in Shetland" for further explanation.
**Sumburgh is the only airport in Scotland with this facility.*

SUMBURGH AIRPORT PASSENGER TRAFFIC

Chart XIV

Year	Total Number of Passengers	Fixed Wing Scheduled Passengers	Fixed Wing Charter Passengers	Helicopter Passengers	Percentage Fixed Wing Scheduled Passengers	Percentage Fixed Wing Charter Passengers	Percentage Helicopter Passengers
1972	70589	45871	13979	10739	65	20	15
1973	94312	53614	21088	19610	57	22	21
1974	158253	67680	52187	38386	43	33	24
1975	186319	75525	64897	45897	41	35	25
1976	250778	78654	84111	88013	31	34	35
1977	404370	84138	168107	152125	21	42	38
1978	685492	90601	323151	271740	13	47	40
1979	622446	89324	279908	253214	14	45	41
1980	608169	82374	268985	256810	14	44	42

Notes:
1. Fixed wing charter and scheduled passenger figures were only available from 1972.
2. Fixed wing charter passengers consist of:
   (a) Oil Rig or platform crew changes
   (b) Oil related construction workers — the majority of these workers ceased to transit through Sumburgh from the end of July 1978 when Scatsta Airport became operational. This accounts for some of the difference between 1978 and 1979 figures.
   (c) Private parties/general usage. The majority of fixed wing charter passengers are in category (a)

Having discussed developments, and the cost of developments at Sumburgh, it is perhaps now appropriate to examine the types of air traffic these developments serve.

(iv) **Types of Air Traffic**

Air traffic at Sumburgh can be divided into the following categories:

(a) Fixed wing aircraft — charter or scheduled;

(b) helicopters.

(a) **Fixed Wing Aircraft — (i) Charter**

Charter is the category where there has been the greatest increase in the use of fixed wing aircraft, with 1971 being the watershed year as far as the change in the pattern of activity of fixed wing operations was concerned. It was also in 1971 that helicopters first became prominent at Sumburgh, and perhaps this might be taken as the start of the 'staging post for helicopter to plane, plane to helicopter movements' period of Sumburgh's existence, when the chartered fixed wing aircraft business began to flourish. The bulk of the charter business is, as one would expect, in the service provided for oil rigs and platform workers connecting with helicopter flights at Sumburgh. Alidair, Dan Air, and Loganair have been the companies most involved in this business.

Until July 1978, when Scatsta Airport became operational, Dan Air also provided charters several times per week to convey oil related construction workers from Sullom Voe via Sumburgh to Aberdeen and Glasgow on the Scottish Mainland.

The smaller private charter plane business has also boomed. The reasons for this are two fold:

(1) Time is of the essence to people, and time can be saved if one can obtain flights to suit one's business schedule.

(2) The cost of charters, even for small groups, has often been less than the cost of seats on scheduled services.

In respect of smaller charter aircraft, most have come from the UK Mainland, visiting Sumburgh only for as long as their clients required them to be in Shetland.

There are also regular small charter aircraft flying every week from Sumburgh to Oslo and Haugesund in Norway, transporting Norwegian oil workers.

(ii) **Scheduled Services**

Scheduled services fly into Sumburgh from Aberdeen, Wick and Kirkwall (see map of air corridors into Sumburgh Airport).

British Airways provides services from Aberdeen and Kirkwall using Hawker Siddeley 748 aircraft (48 seat capacity) for the Aberdeen to Sumburgh route. Viscount (60 seat capacity) are normally used on the route from Kirkwall. Air Ecosse took over the scheduled service from Wick in November 1979 from British Airways, and use a Bandeirante aircraft.

The inter-island scheduled service was operated from Sumburgh from 1970 to May 1980, by Loganair, flying to Tingwall, Scatsta,* Unst, Whalsay, Fetlar and Fair Isle,** using Britten Norman Islanders (up to eight passengers, depending on conditions). This service transferred its base to Tingwall in May 1980.

(b) **Helicopters**

Helicopters began flying from Sumburgh in 1971 when British Airways Helicopters began positioning aircraft there. In 1972, Bristows also started flying out of Sumburgh. A third helicopter company, North Scottish Helicopters, has more recently (1980) completed a new hangar building at Sumburgh and set up a base in March 1980, although its aircraft had used Sumburgh before then.

The service began to Scatsta in 1978.
**The service to Fair Isle began in 1976/77.*

Helicopter traffic has been the area of greatest growth of all traffic through Sumburgh — there being virtually no helicopters using the Airport before 1971 (see Chart XIII of "Sumburgh Airport Usage").

The type of helicopter most used from Sumburgh is the Sikorsky S61N (this usually acccommodates up to 20 passengers depending on freight, etc.) of which there are 16 at Sumburgh in 1981. The S 330 Aerospatiale 'Puma' has been present at Sumburgh over the years, the occasional Sikorsky S 58 'Whirlwind' (which usually accommodates about ten passengers), and the Bolkow 105, a five seater aircraft which uses Sumburgh for servicing but is mainly based offshore for flying short hauls between rigs and platforms. There has been a shortage of both helicopters and helicopter pilots for North Sea operations. For example, helicopters used have been up to 15 years old, having been practically rebuilt with spare parts, because of the limited availability of aircraft. This has not meant any lowering in safety standards — far from it. Helicopters require a very high standard of maintenance as they have so many more moving parts than a conventional fixed wing aircraft and are subject to frequent checks.

New breeds of helicopter are expected to be seen in the Shetland area from 1981. The first of these will probably be the Boeing Vertol 'Chinook'.

Although not a new design — this aircraft has been in military service for a number of years — the Chinook has been adapted for North Sea service, and British Airways Helicopters expect to take delivery of six of these aircraft* which can seat up 44 people, taking probably up to two and a half times the load of the £2 million Sikorsky S 61 N. These big helicopters have a greater range** than the Sikorsky S 61 N and it is intended to fly some of them directly from Aberdeen to the platforms off Shetland. Shell has contracted with British Airways Helicopters*** for this service and has had the helicopter landing decks (helipads) of its platforms specially strengthened in order to take the heavier aircraft. The Chinook will be tried operationally in the North Sea for the first time in 1981, after certification by the CAA. Its performance will then dictate its future usage. It could prove a severe blow to Sumburgh Airport usage as 40% of the oil related traffic through Sumburgh (fixed wing and helicopter) is for Shell's offshore operations. However, British Airways Helicopters have stated their intention to maintain their level of operations at Sumburgh, possibly having some of the Chinooks operating from there.

Another new long range aircraft possibly expected within a year (March 1982) for the North Sea operations is the Aerospatiale Puma 332L, known as the 'Super Tiger'. Bristow Helicopters have placed a £80 million order for 35 of these machines. It is not certain how many will be based in the North Sea area, as the present Bristow fleet of S 61 N aircraft is regarded as having up to ten years of useful life. The Super Tiger is faster than the S 61 N, with a 145 knot cruising speed against 110 knots, and has a much better payload range with capability of carrying 19 passengers directly from Aberdeen to the East Shetland Basin.

A third new long range helicopter, the 12 seater, Sikorsky S 76 'Spirit', is already in operation and is capable of flying directly from Aberdeen to the East Shetland Basin. This aircraft carries only 12 passengers and will probably be used for smaller oilfield developments.

It is probable however, that with the ever increasing cost of helicopter fuel and of helicopter time (due partially to high maintenance costs compared to fixed wing aircraft) that economics will dictate that the new generation of helicopters will continue to operate from Shetland, as did the previous generation. Their longer range will be increasingly useful as the oil search extends beyond latitude 62 degrees North.

The larger carrying capacity of the Chinooks might also increase the tendency to send freight offshore by helicopters, which has been developing in recent years, but it must be emphasised that not all oil related vessels and installations will be able to take Chinooks, and so far only British Airways have ordered them. Up till recently, most oil rigs, fixed platforms, laybarges, and other oil related craft

---

*The cost of the six aircraft and spares is £60 million.

**Their range is over 500 nautical miles.

***British Airways Helicopters have a £78 million contract with Shell for a seven year period.

Shetland schoolchildren on the helideck of Unionoil's Heather platform, 1979, with a Sikorsky S6/N helicopter in the background.
(Photograph courtesy of British Airways Helicopters)

The new Boeing Vertol 234 (Chinook) aircraft. *(Photograph courtesy of British Airways Helicopters)*

have had a 'standard' helideck, with a landing area of at least 63 feet in diameter and unobstructed approaches through an area of 270 degrees. The cost of altering helidecks is high, especially as it could involve structural alterations to the vessel concerned, and not all operators will desire to do this.

Having mentioned the aircraft, it is perhaps worth briefly commenting on the men who fly them. Helicopter pilots until recently were normally recruited from ex-UK armed forces personnel. Due to cutbacks in defence spending and the increasing need for pilots, this source of supply has dried up somewhat, and the companies are now recruiting pilots who have spent four years at flying school. This does not mean a drop in standards, as the requisite number of flying hours and other qualifications for full captain status remain in force.

The helicopter business is expected to continue to expand, whatever the constraints may appear to be. The placing of more platforms on the seabed around Shetland will encourage expansion, more so than the additional exploration activity envisaged. Fixed production platforms give a more regular flight pattern for helicopters than exploration drilling rigs, which may be positioned in the area for only 60 days or so, whilst they drill one exploration well. The installation and completion of facilities on a production platform can take up to one year or more, whilst development drilling from the platform may take up to three years to complete. This gives the helicopter companies the opportunity for more long term contracts, and more regular flight plans than with exploration rigs, although exploration rigs appear to generate more flights when in operation. The Civil Aviation Authority (see Sumburgh Passenger Survey, as noted) estimated an exploration rig worker makes 35 trips per year, compared to 34 trips per year for the production platform worker, and 31 trips per year for the platform construction worker.

2. TINGWALL AIRPORT

The Tingwall strip, six and a half miles north of Lerwick, operated until the Civil Aviation Authority withdrew its licence in 1975. During this time, Tingwall had become one of the main points of call on the inter-islands air service route operated by Loganair, and subsequent to the withdrawal of the licence, Shetland Islands Council took steps to provide a new tarred surface strip on a nearby site.

In October 1976, these newly completed facilities (costing about £250,000) were opened, the airstrip having been relicensed by the Civil Aviation Authority. The facilities include a small terminal building, and a building housing fire services and other amenities. The airport has gradually expanded. Fuelling facilities, fixed wing and helicopter parking, radio masts, and new lighting were installed to cope with the increased usage of the airport.

In the spring of 1979, Loganair began a scheduled service from Tingwall to Edinburgh using De Havilland Twin Otter aircraft (approximately 15 passenger capacity). This three fold increase in traffic can be seen when the first full year of operations — 1980 — is compared to 1978 (the year prior to operations beginning) in Chart XVI.

Before the new strip was constructed a number of sites nearer Lerwick had been considered. For example, in 1973, the possibility of building an airstrip to service Lerwick to be located at the Rova Head in Lerwick Harbour, was rejected by the Lerwick Harbour Trust, the estimated cost of £240,000 being amongst the reasons for this decision.

A further event of considerable importance to Tingwall occurred in July 1980, when Loganair moved their base for scheduled inter-island services from Sumburgh to Tingwall.

Tingwall is proving useful not only for inter-island communications, and ambulance flights, as the nearest airfield to the hospital in Lerwick; it is used in connection with oil related developments at Lerwick Harbour.

3. SCATSTA AIRFIELD

Scatsta aerodrome was initially constructed in the period Spring 1940-mid 1943, in order to provide a shore base for fighter aircraft to protect the Sullom Voe Catalina and Sunderland flying boats, which patrolled the North Sea, and the North Atlantic, during the hostilities current at that time. Two runways were constructed, including perimeter tracks and approaches, in addition to bases for hangars. By mid 1941, the first runway was complete, and this is the same runway which was resurfaced, when the airfield was revived.

The airfield was revived in 1978, after a long period of 'rest', as the ground was de-requisitioned after the war, and had been virtually unused until the Americans built a Loran station at the end of the east-west runway in July, 1968.

EMPLOYMENT AT SUMBURGH AIRPORT

Chart XV

Year	Local[*1]		Total Local	Incomers		Total Incomers	Overall Total (Local & Incomers)
	M	F		M	F		
1969	4	3	7	1	1	2	9
1970	5	4	9	1	2	3	12
1971	7	4	11	5	2	7	18
1972	14	6	20	4	3	7	27
1973	20	6	26	18	4	22	48
1974	35	10	45	21	5	26	75
1975	41	14	55	37	8	45	100
1976	68	26	94	119	12	131	225
1977/78	121	50	171	321	37	358	529
1979/80	172	82	254	422	58	480	734
1980/81	129	88	217	385	48	433	650

*1 Local means resident Shetlanders, usually pre-oil (1969)

2 Part-time workers = 1 full-time worker for the purposes of this count. There were 26 part-time workers included in the 1978 count

3 Numbers in 1979 rose beyond 600 because the new heliport opened and more people were employed as a result.

4 Police and Customs are excluded from these figures as are part-time Bank and Norscot employees

5 In 1980 hotel employment in the immediate area (10 mile radius) encompassed another 56 jobs, 44 of these being held by Shetlanders

6 There was a decline in employment in 1980/81 which was partially due to some of Bristows operations moving to Unst, and Loganair moving to Tingwall

In 1977, following a review of the possible use of Scatsta, prompted by requests from the Department of Energy and the Oil Industry, Sir Frederick Snow and Partners were asked by BP to prepare a feasibility report on the use of Scatsta airfield for regular Dan Air charters, which would utilise the Hawker Siddeley 748 aircraft. The main reason for reviving Scatsta was its proximity to the construction villages of Sullom Voe, and the fact that its use would save a two hour (or more, in bad weather) bus journey from Sumburgh Airport to Sullom Voe, or vice versa, for the workforce making their flight connections.

The Snow Report recommended the upgrading of the existing south-west/north-east runway for takeoff and landing, with the other runway to be used for re-parking of aircraft, and as a site for terminal buildings. The Sullom Voe Association Limited accepted these recommendations, and in October 1977, applied to the Shetland Islands Council for planning permission to implement them. This permission was finally granted in April 1978, subject to certain conditions, of which the most important was that the planning permission granted be valid for a limited period only, expiring on 31st January 1983. Unless a further permission is granted at or before this time, the use of the airfield shall cease and the buildings and other effects will have to be removed as required by the planning authority, and their sites be reinstated.

The main conditions of the lease of the airstrip by the Shetland Islands Council* to BP Petroleum Development Ltd., were that it precluded the use of the airstrip (a) for scheduled air services between the UK Mainland and Shetland; and (b) as a regular interchange point betweeen fixed wing aircraft, and helicopters to service offshore platforms or rigs. (This condition was imposed partly to protect developments at Unst, it being Shetland Islands Council policy to spread oil related developments throughout the islands). Furthermore, the initial lease was for two years only. Planning permission having been granted, work proceeded to reconstitute the airfield as recommended by Snow's report. This entailed the removal of 100,000 cubic metres of hillside to achieve an operational 960 x 23 metres runway, with acceptable side slopes.

Precision Approach Path Indicators were installed on the new airfield, being a landing system which allows a pilot of a landing aircraft to make the maximum use of the available runway length. Other navigational aids at the airfield are standard visual aids, radar,** and a Non-Directional Beacon (NDB). (See Chart XII "Oil Related Airfields in Shetland" for further details).

Also constructed were a new prefabricated terminal building, including control room, fire station, generator room and a workshop building for Dan Air. The management of the airport is contracted to Airwork Limited, who provide all air traffic services, fire services and baggage handling. This development has represented an investment by the oil industry of approximately £2 million.

The first flight took place from the airfield on 24th July, and the airfield was formally opened in August 1978. Three Dan Air HS748 aircraft are positioned at Scatsta, and spend overnight there.

Since the airport opened in 1978, it has become progressively busier, as the Sullom Voe construction workforce built up to a peak at the end of 1980. In 1980 the throughput of passengers averaged about 2600 per week. This is not taking into consideration periods of exceptional activity, as for example, is occasioned by the evacuation of the workforce that has occurred for the Christmas and New Year periods. Over 5000 people were moved out within a few days in December 1980, in a 'Berlin Airlift' type of operation. A special "Evacuation Committee" is established for these occasions, consisting of representatives from Dan Air, Foster Wheeler and BP. The HS748 aircraft fly to Glasgow, with a few to other destinations including Aberdeen.

The return after the holiday period has been less fraught, as some of the workforce have two weeks, some three weeks leave, and no jet shuttle is required. Instead the usual "bus service" operated, with a steady flow of HS748 aircraft carrying 44 passengers at a time.

The "normal" system of travel from Scatsta is for the passengers to have identity numbers, rather than an

---

Shetland Islands Council, as landowners of part of the land on which Scatsta airfield stands, lease this to the BP Pet. Dev. Ltd.; the other portion of Scatsta airfield is owned by Frank Hunter who leases it to the Shetland Islands Council, who in turn sub-lease to the BP Pet. Dev. Ltd.

**Each month there are on average 50 radar approaches made to Scatsta. The Plessey 430 radar enables an aircraft to approach within half a mile of the runway at 430 feet. It costs £5,000 for each flight which cannot get in, so the radar, which cost £¾ million, paid for itself within three months.*

airline ticket, and they present their identity cards to board the flights. The contractors tell their men which day to show up for their flights. Foster Wheeler act as "travel agents" and co-ordinate the requirements of the different contractors, and in turn liaise with Dan Air. Foster Wheeler also organises the passengers, being in radio contact with the construction villages to ensure coach loads of passengers do not depart before planes are ready to receive them, as there is not a large waiting area at Scatsta.

This entire operation at Scatsta was designed around the HS748 aircraft. It was BP's decision as Constructor of the Terminal to make the investment at Scatsta and to use Dan Air,* who were consulted at the design stage of the new airport, in respect of the length and width of the runway, navigational aids, workshop and other such facilities required.

A remarkably smooth operation has resulted. The airport is very rarely closed for weather, as it is well situated, and has excellent snow clearing and de-icing equipment apart from good navigation aids and air traffic control.

4. UNST AIRPORT, BALTASOUND

In the 1960's the Zetland County Council (forerunner of the Shetland Islands Council) pursued an active role in encouraging the development of airfields throughout the Islands. In 1967, the Unst Council of Social Services, in conjunction with the Highlands and Islands Development Board, wrote to the Zetland County Council proposing that an airstrip be built. They had been advised that the Royal Engineers from the British Army, under the auspices of Operation MACC (Military Aid to the Civil Community), might be able to construct an airstrip, providing manpower, technical knowledge and equipment free to the benefitting organisation or people, who had to purchase all the land and materials for the job. The Zetland County Council took the matter in hand, agreed to pay the necessary costs, and acquired land from the Ordale Estate and Mr A. Jamieson of Belmont.

A 6000 ton army logistics ship sailed for Unst in July 1967 carrying the necessary machinery and materials for the project, and the 15 Field Squadron of the 38th Engineers regiment provided a labour force of 50 men. The airfield took two summers to build. In 1967 only 75 percent or so of the earthwork, and 1200 feet of the planned 2400 foot runway was completed, due to adverse weather conditions and the difficult terrain. The airfield was completed in July 1968, and the first aircraft landed in Unst in August. Thereafter the airfield became operational almost immediately for charter flights, and finally obtained its CAA licence for scheduled Loganair flights on 9th April 1970.

From 1976, Unst became the object of increasing oil industry attention as offshore exploration and development activities moved further north and Sumburgh airport became more congested. Several plans for the development of this airstrip were mooted — a £4 million scheme by Shell UK in 1976; a £9 million scheme by Miller Construction and a consortium of companies in 1977 — until finally, in 1978, a group of oil companies, namely BNOC, Chevron, Shell, and Union Oil, came forward to the Shetland Islands Council with a proposal. By this time, (in early 1978), Sumburgh airport had reached saturation point,** with aircraft even diverting to Bergen.

The proposal was for facilities costing £750,000 to be built at the Unst airfield, in order that it could be utilised as a staging post for fixed wing/helicopter oil platform crew changes, using De Havilland Twin Otter aircraft, because of the short runway, and linking up with helicopters in this operation. The Shetland Islands Council gave planning consent for this proposal in April 1978, (at the same time as it agreed to the new heliport scheme for Sumburgh).

Operations began on 11th July 1978 with helicopter flights to and from the Ninian Oilfield, and the platform workers transferring into the Twin Otter (and occasional Britten Norman Trislander) aircraft for flights to Kirkwall or Aberdeen***where onward connections could be made. Bristow Helicopters and Loganair, under contract to Chevron Petroleum (UK) Limited have provided the flying services.

A very considerable upgrading of the airport has been undertaken by Chevron (Ninian Field Operators) on behalf of BNOC, Union Oil, Shell and themselves. A new terminal building and aircraft parking area was built, passenger processing, air traffic control, navigation aids and fire services have all been provided.

Dan Air is part of the Davies and Newman group.
**This was before the new heliport was built at Sumburgh.*
***The Twin Otters also fly into Sumburgh on occasion.*

Since the new facilities at Unst Airport were formally opened by the Rt. Hon. David Howell MP on 3rd September, 1979, improvements have continued to be made.

CHANGING PATTERNS OF OIL RELATED AIR TRAFFIC IN SHETLAND 1978-1980   Chart XVI

I Passenger Movements

Year	Sumburgh	Tingwall	Scatsta	Unst	Total	% of Total at Sumburgh
1978	685492	3666	42601	25711	757470	90.49
1979	622446	8840	124512	76059	831857	74.82
1980	608169	12194	136019	138388	894770	67.96

Comment: Passenger movements increased overall by 18% from 1978 to 1980 but the distribution of traffic changed with developments at the three airports north of Sumburgh.

II Aircraft Movements

Year	Sumburgh	Tingwall	Scatsta	Unst	Total	% of Total at Sumburgh
1978	50666	1442	1786	2691	56585	89.53
1979	47771	2200	6613	6806	63390	75.36
1980	41307	2989	6404	11314	62014	66.6

Comment: Aircraft Movements increased by 9.59% from 1978-1980 and like passenger movements, distribution of traffic changed.

Changes are expected again when the construction phase runs down at Sullom Voe and traffic at Scatsta decreases.

High intensity runway lights, low intensity taxiway lights, apron flood lights and Precision Approach Path Indicators (PAPI's) were installed in September 1979, and a cloud base measuring searchlight in October 1980. An extended over run/under run has been provided to the west of the runway, and various improvements made in the drainage and landscaping of the safety strip and dispersal areas.

In May 1980 Bristow Helicopters Limited provided a £1 million hangar to house their S61N helicopters and moved their Chevron operation to Unst. Subsequently the monthly average operations from Unst increased to approximately a total of 1000 fixed wing and helicopter aircraft movements and 12,000 passenger movements.* One result of this acitivity was that during the latter half of 1980, oil company personnel were billeted at the Hagdale Lodge Hotel, Baltasound, and commuted daily to the Ninian (Chevron) and Thistle (BNOC) fields. For several months Unst thus gained an extra 60 temporary residents.

Changes are expected in 1981, when it is anticipated that the four engined, 50 seat, De Havilland Company Dash 7 aircraft, will be used by Chevron to replace all or some of the Twin Otters used to date.

---

*For the growth of passenger movements at Unst, and other airports see Chart XVI "Changing Patterns of Oil Related Traffic in Shetland 1978-1980".

By the beginning of 1981, Unst airport activities provided full time employment for 68 people, of whom approximately 40 percent were local. This is in contrast to the time before oil related developments occurred at the airport (pre 1978), when the airport provided no full time employment.

AIR SERVICES EMPLOYMENT IN SHETLAND (as at February 1981)   **Chart XVII**

Airport	Total Employees	Male	Female	% Local	% Women
Sumburgh	650	514	136	33	21
Tingwall	12	9	3	92	25
Scatsta	42	36	6	33	14
Unst	68	57	11	37	16
Total	772	616	156	49 (average)	19 (average)

Note: (1) % figures are rounded to nearest whole number

(2) Part-time jobs are included in these figures

II THE DEVELOPMENT AND INTEGRATION OF FLIGHT ADVISORY AND CONTROL SERVICES TO COPE WITH AIR TRAFFIC GENERATED BOTH ONSHORE AND OFFSHORE

(a) **Onshore Shetland**

Air traffic control services were first operated in Shetland at Sumburgh. The Civil Aviation Authority provided a service for aircraft using the airport. From 1973, this service covered Sumburgh airport's responsibility up to a 40 mile radius from the airport for fixed wing aircraft and up to a 25 mile radius for helicopters, which thereafter transferred to the Scottish "Offshore North" sector of the National Air Traffic Services (NATS) system operated from Prestwick in Scotland. The Civil Aviation Authority have since extended their NATS system to cover most of the UK Sector of the North Sea, and although there were new radio communications links installed at Scousburgh, and an extension of the Flight Information Service from Prestwick, further improvements were still requested by the operating companies in the East Shetland Basin, especially for areas at a distance from Sumburgh.

In 1974, a near collision between two S 61 N helicopters in poor visibility off Shetland led the helicopter companies to establish their own Flight Advisory Service which became operational in April 1975. This lasted nearly three years until 1978 when the NATS facilities were extended. The Flight Advisory Service operated by the oil companies had extended beyond 100 miles from Sumburgh to the most distant Fields then being explored.

The main objective of this service was to ensure that the helicopters maintained a safe distance from each other. This can still provide some anxiety to helicopter pilots as congestion increases in the area and activity moves further north and west.

Since 1978 developments occurred which have increased the complexity of the problem. As stated Tingwall, Scatsta, and the Unst Airports have very greatly increased their traffic.* There is no Air Traffic Control service at Tingwall, but aircraft flying to Tingwall 'overfly' Sumburgh, and utilise the NATS Service provided there, even if they are not landing there. Similarly, aircraft flying to Scatsta and Unst require to 'overfly' Sumburgh. The Air Traffic Control at Scatsta deals mainly with the traffic there, but again has aircraft 'Overflying' — this time to Unst.

*See Chart XVI "Changing Patterns of Oil Related Air Traffic in Shetland 1978-1980"

AIR ROUTES : NORTHERN NORTH SEA

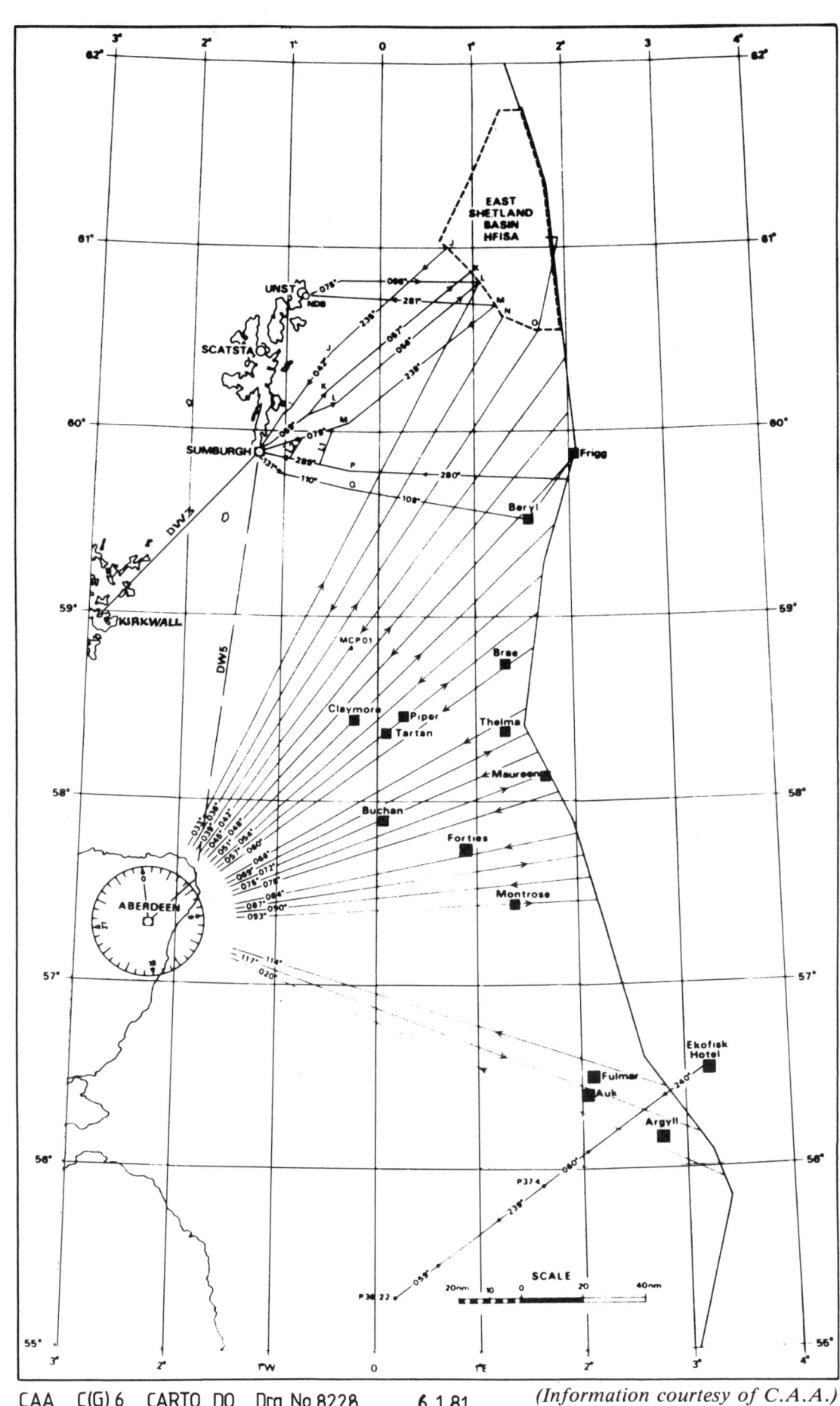

CAA C(G)6 CARTO DO Drg No 8228 6 1 81 (Information courtesy of C.A.A.)

At Unst the situation for the airport would have been impossible without the help and co-operation of 'Shetland Radar' operated by members of the Royal Air Force who, since 1978, have provided an area radar service* twelve hours a day, seven days a week, mainly for helicopters which are flying from Shetland (this includes Sumburgh and Unst)** to the oilfields east and north, this service helps insure there is good flight advisory information available to the helicopters in order that there are no conflictions between aircraft flying out to and from the oilfields.

AIR ROUTES: NORTHERN NORTH SEA

Originally there were two Air Traffic Advisory Routes:

(i)   Delta White 3 (DW3) — Sumburgh via Kirkwall and Wick to Aberdeen.

(ii)   Delta White 5 (DW5) — Sumburgh direct to Aberdeen.

(Delta White 5 West (DW5W) was introduced as an alternative route to avoid military danger areas when active, The danger areas have now been re-aligned and DW5W no longer exists).

Increased traffic to the East Shetland Basin meant that a helicopter route structure had to be devised and this has been developed over the last two years (it was introduced about Spring 1979). Helicopter routes from Unst were added when operations at that airfield increased.

When new radar equipment became available at Sumburgh the Offshore Flight Information Service operated from Scottish Air Traffic Control Centre at Prestwick was replaced by the radar advisory service and the airspace was divided between Shetland Radar, Sumburgh ATC, Highland Radar and Aberdeen ATC.

An increase in traffic overflying Sumburgh to Scatsta, Unst and Tingwall presented further air traffic problems and a Radar Advisory Service Area, controlled by Sumburgh, was introduced in September 1980.

To give added protection to aircraft operating into Sumburgh, a Special Rules Zone/Special Rules Area is to be introduced and plans are in progress to upgrade the old advisory routes to full airway status.

Shetland Radar, sited at RAF Saxa Vord, use a military radar to operate a safe lateral separation system, in addition to a system of separation based on time and altitude, where one helicopter is no closer than five minutes flying time and no nearer than 500 feet in height from the one ahead or behind. This system has been possible because most of the helicopters have been flying at about 130 miles per hour except the Aerospatiale Pumas which fly 40 miles faster and are allowed to overtake.

The helicopters navigate by flying along magnetic radials, and are monitored every ten minutes. If they do not report their position a warning device goes off in the cockpit. This practice of reporting every ten minutes allows the location of a ditched helicopter to be fixed within 20 miles or less***(the helicopters have floats attached to their undercarriages so that, if necessary they can land on the sea). This procedural service is backed up by radar. Each helicopter carries a personal identity code which it 'squawks' when picked up by the radar, the aircraft showing up as dots on the screen with small tadpole like tails indicating in which direction they are moving. All three radar advisory and control services, at Sumburgh, Scatsta, and Unst, work in co-ordination.

Shetland Radar provides an advisory service to aircraft from a 25 to 90 nautical mile radius North East from Sumburgh, with aircraft calling in at 30, 50, 70 and 90 nautical miles out. The aircraft fly out on one radial and back on another to avoid confliction. A measure of how active this service has been is that in its first two years of operation, from 1978 to 1980 it monitored 60,000 aircraft movements.

An area radar service covers an area of airspace as opposed to airfield approaches
**70% of the Shetland traffic comes from Sumburgh*
***For ditched helicopters and others offshore there is a helicopter search and rescue service, which was started in June 1979 from the Brent Field by Shell, in conjunction with Bristows Helicopters.*

NORTHERN NORTH SEA RADAR SERVICE AREA INFORMATION AND ASSOCIATED RADAR SERVICE AREAS

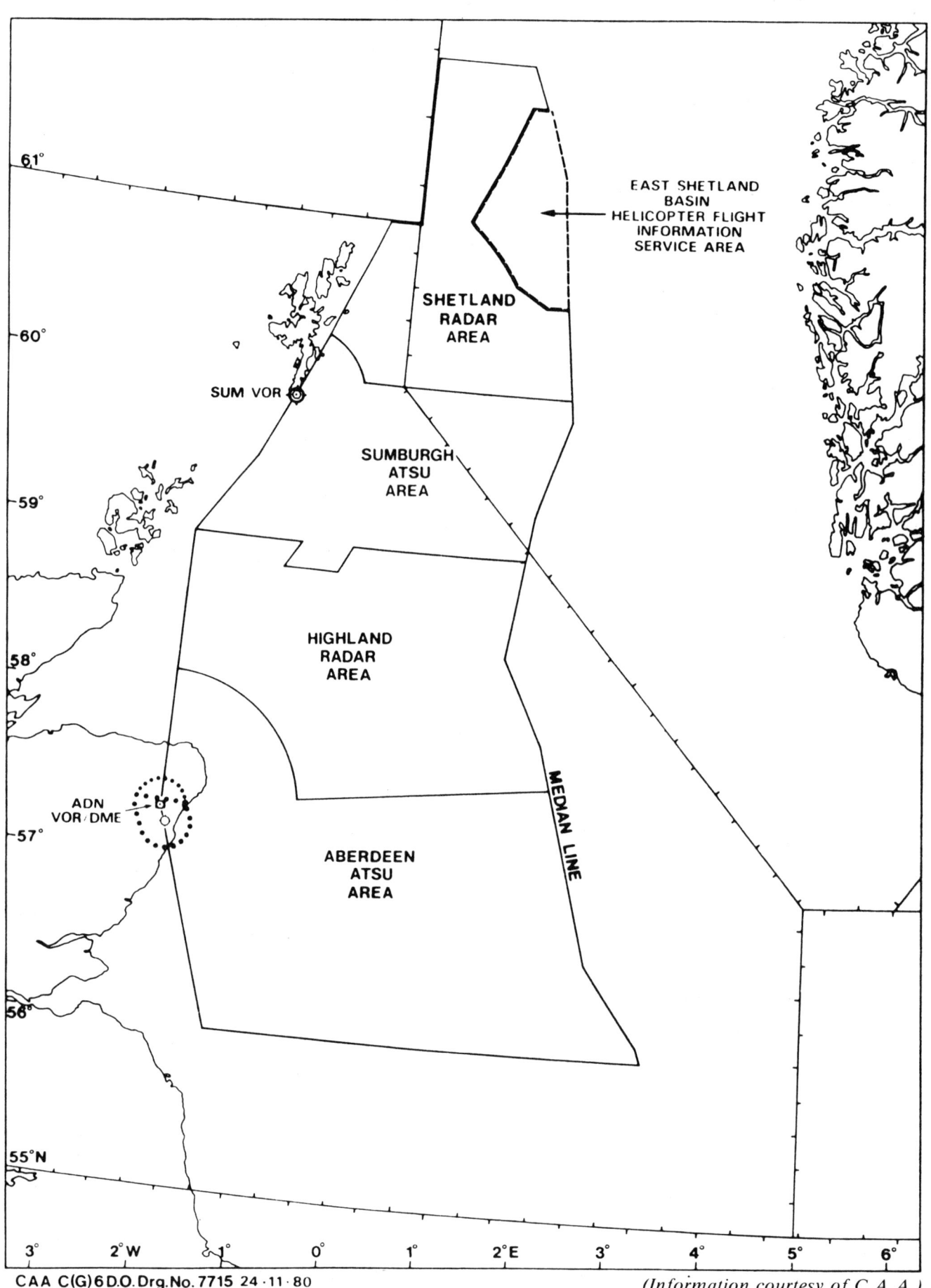

RADAR & NAVIGATIONAL AIDS

Table IX

In order to control the increased air traffic operating to the airfields in Shetland and in transit to the East Shetland Basin a fairly comprehensive system of radars and navigational aids is required. The air traffic radars now available in Shetland are as follows:

Sumburgh	Long range surveillance radar (120 nautical miles); Long range secondary surveillance radar (150 nautical miles); Short range surveillance approach radar (30 nautical miles range) used primarily for radar controlled approaches to the airfield.
Saxa Vord (Shetland Radar)	Long range surveillance radar; Long range secondary surveillance radar.
Scatsta	Short range surveillance and approach radar used primarily for radar controlled approaches to the airfield.

These radars are supplemented by a number of radio navigation aids. They are as follows:

VOR/DME	Very High frequency omnidirectional Range (VOR) and co-located Distance Measuring Equipment (DME). Very accurate navigational aid used for area navigation and instrument approach procedures.
TACAN (Saxa Vord)	Military equivalent of VOR/DME — used for area navigation.
NDB (Sumburgh) (Scatsta) (Unst) (Tingwall)	Non-directional Beacon — frequency navigation aid not as accurate as VOR/DME. Is usually used as an airfield navigation aid and can be used for let down procedures (not at Tingwall). At Sumburgh it is used as standby in case the VOR/DME is out of service.
DECCA	Low frequency area navigation aid. Very accurate. Extensively used by helicopters operating to and from the East Shetland Basin.

AIR TRAFFIC CONTROL

Sumburgh	Air Traffic Control at Sumburgh is provided by the National Air Traffic Services using civil (CAA) controllers. The Unit provides the following services: Aerodrome Control; Approach Control; Approach Radar Control (including radar approaches to ½ nautical mile from touch down); Radar Advisory Service (within designated areas); Flight Information Service; Alerting Service.
Scatsta	Air Traffic Control at Scatsta is provided by Airwork Ltd. The Unit provides the following services: Aerodrome Control; Approach Control; Approach Radar Control (including radar approaches to ½ nautical mile from touch down); Flight Information Service; Alerting Service.
Unst	Air Trafffic Control at Unst is provided by Airwork Ltd. The Unit provides the following services: Aerodrome Control; Approach Control; Flight Information Service; Alerting Service.
Tingwall	At present there is no Air Traffic Control Service at Tingwall. A Flight Information Service is available on request.
Shetland Radar	Air Traffic Control at Saxa Vord is provided by the National Air Traffic Services using military controllers. The Unit provides the following services; Area Radar Control and Advisory Service; Flight Information Service; Alerting Service.

New radar equipment at Sumburgh has led to further developments in the co-ordination of air traffic advisory and control services. The offshore Flight Information Service, which was operated from the Scottish Air Traffic Control Centre at Prestwick, has been replaced by a radar advisory service and airspace in and around Shetland is divided between Shetland Radar, Sumburgh Air Traffic Control, Highland Radar and Aberdeen Air Traffic Control. A Radar Advisory Service Area* controlled by Sumburgh was introduced in September 1980 to cope with the increase in Traffic overflying Sumburgh to Tingwall, Scatsta, and Unst. Further, to give added protection to aircraft operating into Sumburgh, a Special Rules Zone/Special Rules Area is to be introduced, and plans are in progress to upgrade the old advisory routes to full airway status.**

(b) **Offshore Shetland (Beyond 90 nautical miles from Sumburgh)**

A Helicopter Flight Information Service Area (HFISA) exists which is operated by Shell UK Exploration and Production for the oil industry, under a contract to International Air Radio. This is a helicopter control and co-ordination service for the East Shetland Basin area (see map) and covers the Brent, Cormorant, Dunlin, Heather, Hutton, Ninian and Thistle Oilfields.

The service, called East Shetland Approach, stemmed from discussions between offshore operators in the area, and the helicopter companies, which resulted in the Civil Aviation Authority issuing a NOTAM (notice to airmen), an official document outlining the procedure for all aircraft entering and leaving the designated area. Prior to the HFISA becoming operational in January 1979 Shell had, since its first Brent platform floated out 18 months previously, provided their own Brent Information Service for their own platforms, using fully qualified and licenced air traffic controllers working on a system of flight paths similar to that required of a busy airport. For example, in the peak month of July 1978, the Brent Area alone handled 21,653 air traffic movements — this compared with the 26,000 at Heathrow, the world's busiest airport. Since the other fields in the East Shetland Basin came into this control system, movements have doubled.

The control now operated is divided into two distinct parts. The "East Shetland Approach" is for aircraft travelling to, from and through the area. The "Brent Approach" is a local service for Shell's own installations.

Aircraft heading to an installation in the East Shetland area first contact East Shetland Approach and are then handed over to the relevant field approach as they near their destination. On the way out, control is handed from the field to East Shetland Approach.

These procedures take care of one of the problems which arose in 1978, when aircraft from Unst and Sumburgh were flying on possibly conflicting paths. Ninian bound helicopters from Unst which crossed the flight paths of Brent bound aircraft from Sumburgh now join and leave the East Shetland flight paths nearer to Shetland under observation by Shetland Radar.

The East Shetland Approach service operates from the Cormorant Field, where Shell also have a Brent logistics centre which liaises with the air traffic controllers and decides which passengers or cargo will be carried to which destinations, in aircraft going to and from Shell installations, or on the infield "shuttle".

Further developments yet are anticipated in the sphere of air traffic control services in and around Shetland as oil and gas exploration and production proceeds both to the North and West of the Islands.

---

*See map
**An airway is controlled airspace

EAST SHETLAND BASIN AIR ROUTE STRUCTURE

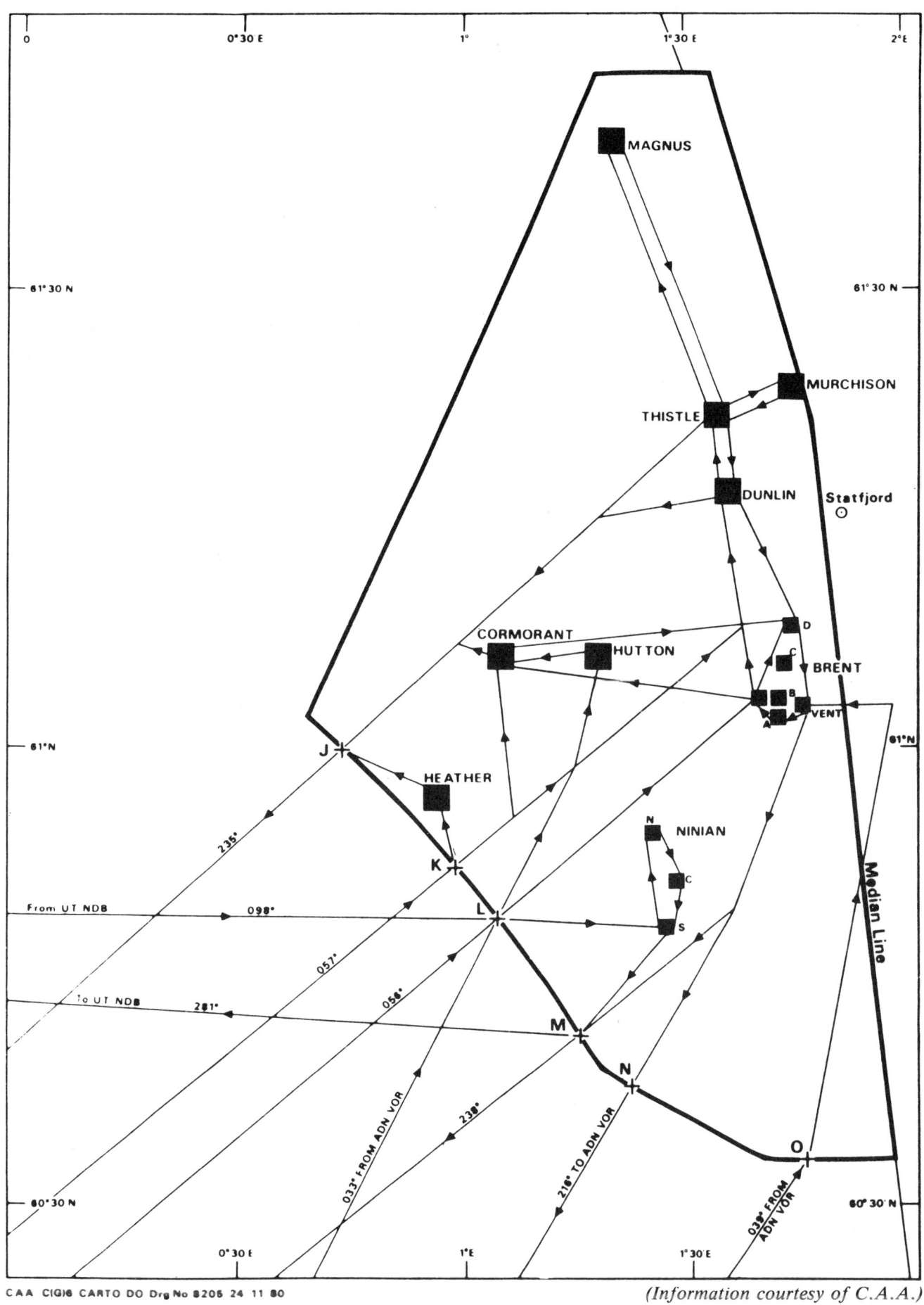

(Information courtesy of C.A.A.)

Chapter Five

The Sullom Voe Oil Terminal

BACKGROUND

It became clear at an early stage in the exploration for and development of oil and associated gas offshore Shetland that the most practical method of transportation would be by sending crude oil and its associated gas by subsea pipelines to a terminal onshore at the nearest landfall. This Terminal would be required to stabilise, store and ship the oil and gas to refineries elsewhere.

In early 1972, therefore, the Zetland County Council appointed the engineering consultants Transport Research Limited to carry out a study of the locations in the Shetland Islands which might be suitable for a major oil terminal. This study, completed in July 1972, identified the Sullom Voe area as having the greatest potential for this kind of development. Perhaps the most important point that was made in this study was the promotion of the idea of the containment of such an oil related development within one area. The Interim County Development Plan, which was finally sanctioned by the Zetland County Council and submitted to the Secretary of State for Scotland in March 1973, ratified this recommendation (see Chapter Two for further details on planning aspects).

Further study of the subject, for example, the Master Plan prepared by Livesey and Henderson,* eventually pared down the somewhat widespread Sullom Voe area originally identified by Transport Research Limited into a fairly intensive development area around Calback Ness (see layout of 'The Sullom Voe Terminal').

A period of extensive consultation began between the Zetland County Council, the UK Government, and the Oil Industry, in order that a means of managing these impending developments be agreed. From the passing of the Shetland County Council Act in April 1974, the way ahead became clearer. The Council and the Oil Industry had already set up a Joint Planning Group to consider the requirements of a Crude Oil Terminal in Shetland. In February 1974, a contract for the preparation of the Shetland Crude Oil Terminal Overall Development Plan was awarded to Foster Wheeler Limited, in association with Sir Alexander Gibb & Partners. This plan was published in April 1974 and outlined the stages of development needed for a single Oil Terminal, receiving a total potential throughput of three million barrels of oil per day from three pipelines. Terminal design considerations were its primary aim.

Terminal Design Considerations

The Council had already taken steps to acquire the land on which the Terminal was to be sited; the design and intended mode of operation of the Terminal now became a major concern to the Council. Several reasons existed for this, one of the most important being that, under the terms of the 1974 Act, the Zetland County Council was both the Local Authority and Harbour Authority. The recent reform of local Government in Scotland** had changed the local authority's responsibility for the enforcement of petroleum legislation for the safekeeping of petroleum spirit — thus, the enforcement of the Petroleum*** (Consolidation) Act 1928, had come to rest with either a regional council or an island council, provided that in any harbour authority, whether or not that harbour is within the jurisdiction of a local authority, the harbour authority shall be the local authority for granting petroleum spirit licences to the exclusion of any other local authority.

---

*This study, which was completed in September 1973 examined six potential Crude Oil Terminal sites. Four of these were soon eliminated and of the two remaining Sullom Voe and Swarbacks Minn, the former was felt to have the advantage.
**See Local Government (Scotland) Act 1973.
***Petroleum includes crude petroleum, oil made from petroleum or from coal, shale, peat or other bituminous substances and other products.

The ZCC Act thus placed the Council, as both the local and the harbour authority, in an ideal situation in having a common administrative link for the control of developments* (other legislation such as the Town and Country Planning Act 1947, and the Health and Safety at Work Act 1974 also affected the implementation of the Petroleum (Consolidation) Act.

The Petroleum (Consolidation) Act 1928 requires that, subject to certain exemptions, petroleum products which have a flashpoint below 73 degrees Fahrenheit shall not be kept, unless a licence is in force authorising the keeping thereof. The petroleum-spirit** is kept in accordance with such conditions as may be attached to the licence.

There is no prescribed form of licence and it is up to each licensing authority to decide what conditions it wishes to attach to the licences. A licensing authority may attach to any petroleum-spirit licence:—

> "Such conditions as they think expedient, as to the mode of storage, the nature and situation of the premises in which, and the nature of goods with which petroleum-spirit is to be stored, the facilities for the testing of petroleum-spirit from time to time, and generally as to the safe-keeping of petroleum-spirit".***

The principal areas of importance to the Zetland County Council (and its Successor, the Shetland Islands Council****) in the design of the Terminal (and not necessarily in order of importance) from the view point of its petroleum licensing responsibilities**** were:—

(a) The relationship between the principal hazard areas, e.g. crude storage, liquid petroleum gas (LPG) storage, and the harbour;

(b) Means of preventing tank spillage or overfill situations including the safety arrangements proposed e.g. high level tank alarms, and cut offs, communications, surge facilities and terminal shut down procedures;

(c) What happens in a spillage situation with respect to bunds, drainage systems, etc.;

(d) Firefighting arrangements and general emergency procedures, including manning levels;

(e) Means of preventing unauthorised access to petroleum-spirit e.g. fences, terminal security, etc.;

(f) Safety of personnel;

(g) Safety of adjoining properties including means of prevention of spillage into the harbour area;

(h) General provisions relating to the safe keeping of petroleum-spirit.

The UK Government through the Health and Safety Executive also proved helpful at this time to the Council as a source of specialist advice on the storage of petroleum.***** An example of design consideration which the

---

*A comparison can be seen on the River Forth, where a substantial part of the tank farm adjacent to the River Forth at Grangemouth is licensed by the Forth Ports Authority whilst the bulk of the products in the main refinery complex alongside are licensed by the Central Regional Council.

**Petroleum-spirit means such petroleum as when tested in manner set forth in part II of second schedule of Petroleum (Consolidation) Act gives off an inflammable vapour at a temperature of less than 73 degrees Fahrenheit.

***See Petroleum (Consolidation) Act 1928, Section 2(3).

****The last working session of the Zetland County Council took place on 22nd April 1975, prior to the formation in May 1975 of the Shetland Islands Council, under the reformation of local government in Scotland (see appendix to Chapter Two).

****Many of these points are either standard industry practice, or covered by statute.

*****Petroleum appeals against refusal of licensing authorities to grant licences are heard by an Explosives Inspector who thereafter reports to the Secretary of State for Employment to whom the Health and Safety Executive is responsible — hence the Health and Safety Executive offer help and advice to try and avoid an 'appeal' situation arising.

Council was concerned with was the question of the storage of the oil and gas, which arose in 1974. The Council was anxious that there be an impervious membrane in all licensed bunded* areas to prevent seepage into the ground,** in the event of spillage of oil occurring. The word 'bund' is not used in the Home office code*** by which licensing authorities abide, but it does recommend that all above ground petroleum-spirit tanks should be completely surrounded by a fire wall, unless the topography of the surrounding area is such that either naturally or by construction, spillages can be directed away to a safe impounding basin. In addition, the model conditions of licence clearly state that due precautions shall be taken "to prevent the escape of petroleum-spirit into any public drain, sewer, harbour, river or canal". The Council therefore desired to marry these two recommendations and required that the bund be not only capable of restricting the spread of fire, but also of holding any spilled product, and reserved the right to test the enclosure.

No direction was given as to the choice of material, and it was the oil industry themselves who selected a concrete lining, since they could not guarantee the integrity of the bund walls using only naturally available materials. Having accepted this principle, it was thereafter necessary to strike a balance between the holding capacity of the bund, the height of the bund walls, and the surface area of the enclosure. These three areas of interest conflict, in so far as one wants total containment, not too high walls for structural and safety reasons, and as small a surface area as possible for fire fighting purposes. Considering the rapid tank product turnover rates at the Terminal, the design concept that evolved of two storage tanks in one bund with a capacity equal to 110 per cent of one full tank was a reasonable compromise, and is both safe, and safeguards the environment.

Design considerations resulting from the Council's responsibility under the Petroleum (Consolidation) Act 1928, were consolidated in the design of the Terminal by the Oil Industry. This design complied with the current (1974) Home Office,* and Institute of Petroleum codes.****

Environmental considerations were also important in the design context, for example:—

(a) where possible tanks and other Terminal facilities were located to use land contours to provide screening;

(b) Large items (e.g. the Power Station) were to be painted in neutral colours;

(c) The installation was to be designed to avoid any oil spillage during operations with bunding and catchment systems provided to retain any accidental oil spillage within the Terminal for disposal under controlled conditions;

(d) Standard designs of floating roof tanks which are used for crude oil storage, were modified by the Industry to withstand Shetland wind loadings; subsequently further stiffening was added at the request of the Shetland Islands Council;

(e) Ballast water from tankers and water drainage from oil areas were to be treated in separators and catchment areas so that effluent water should have an agreed minimal oil content.***** Effluent water disposal was made the subject of a special study.

In the following areas the industry substantially exceeded legal requirements:—

(a) Portable facilities were to be provided to deal with accidental marine oil spillages;

(b) Valves were to be provided in crude oil lines at jetties to minimise the leakage of oil in the event of breakage of lines and equipment caused by marine collision or similar accident;

(c) Safety measures incorporated in standard Oil Industry design practices to minimise fire risk, and in addition a comprehensively equipped fire fighting unit was established and manned at all times.

---

*The word bund is derived from the Hindi word meaning bank.

**See "Model Code of Principles of Construction and Licensing Conditions for Petroleum Installation — Part II Distribution Depots and Major Installations" published by Home Office 1968.

***The Home Office is a Department of the UK Government.

****The Institute of Petroleum formulates safe codes and standards of practice in respect of technical matters concerning the oil industry.

*****This was in accordance with normal industry practice.

Construction at Calback Ness, September 1980. Aerial view looking south across crude oil tankage towards jetties. (This shows bunding arrangements). *(Photograph courtesy of BP)*

The above noted considerations, incorporated in a 'Shetland Crude Oil Terminal Overall Development Plan', were agreed between the Zetland County Council and the Oil Industry*. The Council suggested that this Overall Development Plan might be considered for outline planning permission, and obtained the consent of the Scottish Office Development Department (a Department of the Scottish Office under the Secretary of State for Scotland), to do so. Thereafter the Oil Industry could make detailed planning applications (on the basis of this plan being the outline to the Council).**

Not specified in the Overall Development Plan*** were the number of storage tanks which would be required for the crude oil as a cavern storage design was being investigated, as environmentally more acceptable to the Shetland Islands Council. The consultants "Geostock" were commissioned by the Council to ascertain whether cavern storage might be feasible, and Geostock retained the Company Hagconsult to carry out a borehole investigation to this end. It was hoped that it might have been possible to proceed with the design and construction of 1.5 million net cubic metres of underground storage. By mid 1975, Hagconsult and Geostock had completed their work on the site, the results of which were to indicate that the rock conditions were possible but not ideal for cavern storage. Further research revealed it would have been substantially more expensive to adopt such a design. Consequently in 1976 it was agreed that crude oil storage would be in above ground tanks, following the guidelines laid down in the Overall Development Plan.

Terminal Construction and Operational Considerations

Disturbance

Apart from design considerations, the Zetland County Council (and its successor, the Shetland Islands Council) was obviously concerned about the effects on the local community of the massive construction that was to be required at Sullom Voe. Therefore before construction of the Terminal commenced a Disturbance Agreement was made between the Council and the Oil Industry, which was signed by three oil companies in 1974 (on behalf of others with oil discoveries in the East Shetland Basin).

This agreement guaranteed the Council payments for the disturbance that oil developments would bring. The first instalment was paid at the signing of the agreement on 12th July 1974.**** The signing of this agreement eased the way for the construction to begin of the multi user crude oil storage, treatment and shipping facilities at Sullom Voe.

Planning permission was given for the Preliminary Works contract in September 1974 but it was clear, that for the design and construction (of what was to become Europe's largest oil transhipment Terminal), considerable consultation would be required if planning permissions were to progress smoothly. Furthermore, from the powers bequeathed by the ZCC 1974 Act, the Council was able to share in the profitability of Oil Industry operations to the general benefit of Shetland.

As a result, it was decided to create a joint Council/Oil Industry organisation for the management of the Terminal and Harbour developments. This embryo organisation began to function about the time the ZCC Act became law (in April 1974), eventually emerging and becoming incorporated as the Sullom Voe Association Limited,***** on the 23rd June 1975, as a 50/50 joint venture company between Shetland Islands Council and the Oil Industry. The Council, as the 'A' members had two Directors, and the Oil Industry, the 'B' members had two Directors from Shell and BP, as operators of the two pipelines scheduled to bring the oil into the Sullom Voe Terminal — the Brent and Ninian Pipeline Systems. The Chairman has no casting vote.

See 'Shetland Crude Oil Terminal Overall Development Plan' by Foster Wheeler Limited, in association with Sir Alexander Gibb & Partners for the Joint Planning Group of the (now inactive) Oil Liaison Committee. This Oil Development Plan was to be capable of alteration by the agreement of the Sullom Voe Association members as the project took shape.
**Planning procedures can be in two stages: "Outline" planning permission which is for agreement in principle to a project sought; "detailed" or full planning permission, based on more detailed facts if "outline" can be obtained.*
***The Overall Development Plan and subsequent studies were paid for by the Oil Industry.*
****These monies have been placed in a Charitable Trust which stood at about £18 million in February 1981.*
*****The Sullom Voe Association is a non-profit making organisation.*

SULLOM VOE OIL TERMINAL — ORGANISATIONAL RELATIONSHIPS

- PIPELINE OPERATORS MANAGEMENT COMMITTEE
- BRENT SYSTEM PARTICIPANTS (20 COMPANIES, 4 LICENCE GROUPS)
- NINIAN SYSTEM PARTICIPANTS (13 COMPANIES, 3 LICENCE GROUPS)
- MANAGER SULLOM VOE TERMINAL PROJECT
- SULLOM VOE TERMINAL CONSTRUCTION (B.P.)
- SULLOM VOE TERMINAL OPERATOR
- JOINT EMPLOYMENT MONITORING GROUP
- SULLOM VOE OIL SPILL ADVISORY GROUP
- SULLOM VOE TERMINAL TECHNICAL WORKING GROUP
- SHETLAND OIL TERMINAL ENV'L ADVISORY GROUP
- LAND SUBLEASE AGREEMENT
- S.V.A. AGREEMENT AND CONSTITUTION

S.V.A. SULLOM VOE ASSOCIATION LTD.
PARTICIPATION
50% SHETLAND ISLANDS COUNCIL
50% OIL COMPANIES FROM BRENT AND NINIAN PIPELINE SYSTEMS

- HOUSING SERVICES AND INFRA-STRUCTURE — S.I.C. AS LOCAL AUTHORITY PROVIDING FOR HOUSING & INFRA-STRUCTURE NEEDS
- CATERING SERVICES — S.I.C. IN JOINT VENTURE WITH *GRAND METROPOLITAN HOTELS* TO FORM *GRANDMET SHETLAND LTD.*
- TOWING SERVICES — S.I.C. IN JOINT VENTURE WITH *CLYDE CORY* SHIPPING COMPANIES TO FORM *SHETLAND TOWAGE LTD.*
- LAND LEASE AGREEMENT — S.I.C. AS LANDLORD
- PORTS AND HARBOURS AGREEMENT — S.I.C. AS PORT AND HARBOUR AUTHORITY
- S.I.C. AS LOCAL GOVERNMENT

For the construction of the Terminal, the Council and Oil Industry agreed that the appointment of one representative Company or Constructor to act on behalf of the thirty or so oil companies using the Terminal would help to ease management problems.

Shell, which was responsible for co-ordinating the original design and early construction of the Terminal, informed the Council that it was intended to transfer this responsibility to BP, as Constructor. Shell acted as Constructor until about the end of 1975, when BP formally took over this role. (The Terminal design was not basically altered or delayed as a result of this transfer).

The Sullom Voe Association (SVA) established a Technical Working Group (TWG) to discuss and agree technical matters pertaining to the design, construction, maintenance and operation of the Terminal. This eased planning procedures, as all future planning applications were to go to the Technical Working Group and their comments were to be sought before the application was processed further.

The applications had then to be presented to and agreed by the SVA Board before being submitted to the Shetland Islands Council by the Terminal Constructor, on behalf of the SVA. From its establishment in May 1975, Shetland Islands Council agreed that all planning applications would be considered by the full Council which would act as the Planning Committee.

The Sullom Voe Association also established various other Committees or Groups to achieve a Joint Council/Oil industry management of the development (see Organigram attached) of "The Sullom Voe Association").

A complete network of organisational relationships has built up around the Sullom Voe Association. In order to understand these relationships, it is perhaps best to first of all consider the development which precipitated them — the Sullom Voe Terminal.

The Sullom Voe Terminal — the Construction and Operation

A major industrial area centres on the Sullom Voe Terminal at Calback Ness, and its ancillary services at Scatsta. To understand the operation and the likely size of the terminal constructed, one has first of all to acquire some idea of the amount of oil that is likely to be handled from the pipelines feeding into it.

Although pipeline facilities within the terminal were designed with the possibility that there might ultimately be three pipelines, initially, the oil from two 36 inch diameter pipelines has been accommodated:

(i) The Brent System pipeline with a nominal capacity of 1,050,000 barrels per day of crude oil, from the Brent, Cormorant area, Dunlin, Hutton, North West Hutton, Murchison and Thistle Fields;

(ii) The Ninian System pipeline, with a nominal capacity of 950,000 barrels of crude oil per day from Heather, Magnus and Ninian Fields (see map).

Originally, a Terminal to handle 300,000 barrels per day was proposed. As a result of subsequent offshore discoveries the scale of the project has grown, and the Terminal is currently designed to handle a 1.41 million barrels per day throughput, although the pipelines have a nominal capacity of two million barrels per day.

It is possible that a third pipeline from the Clair Field to the West of Shetland will come into Sullom Voe later in this decade, (the Clair Field being approximately 34 miles West North West of Sullom Voe).

It is, however, the Brent and Ninian pipeline, and the other facilities presently existing at Sullom Voe, with which we are more immediately concerned.

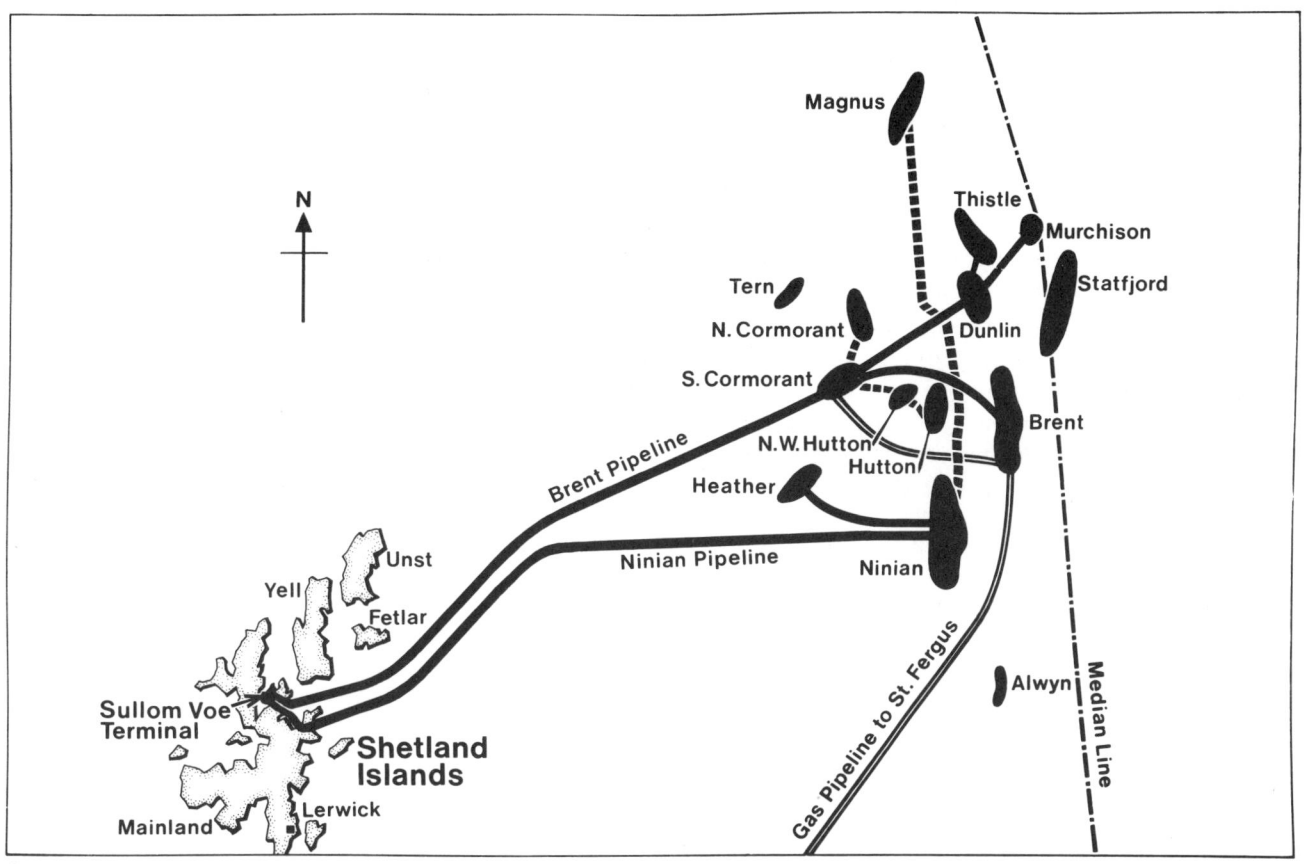

Sea routes of the two 36-inch diameter pipelines from the Ninian and Brent Group oilfields to the Shetland Islands.

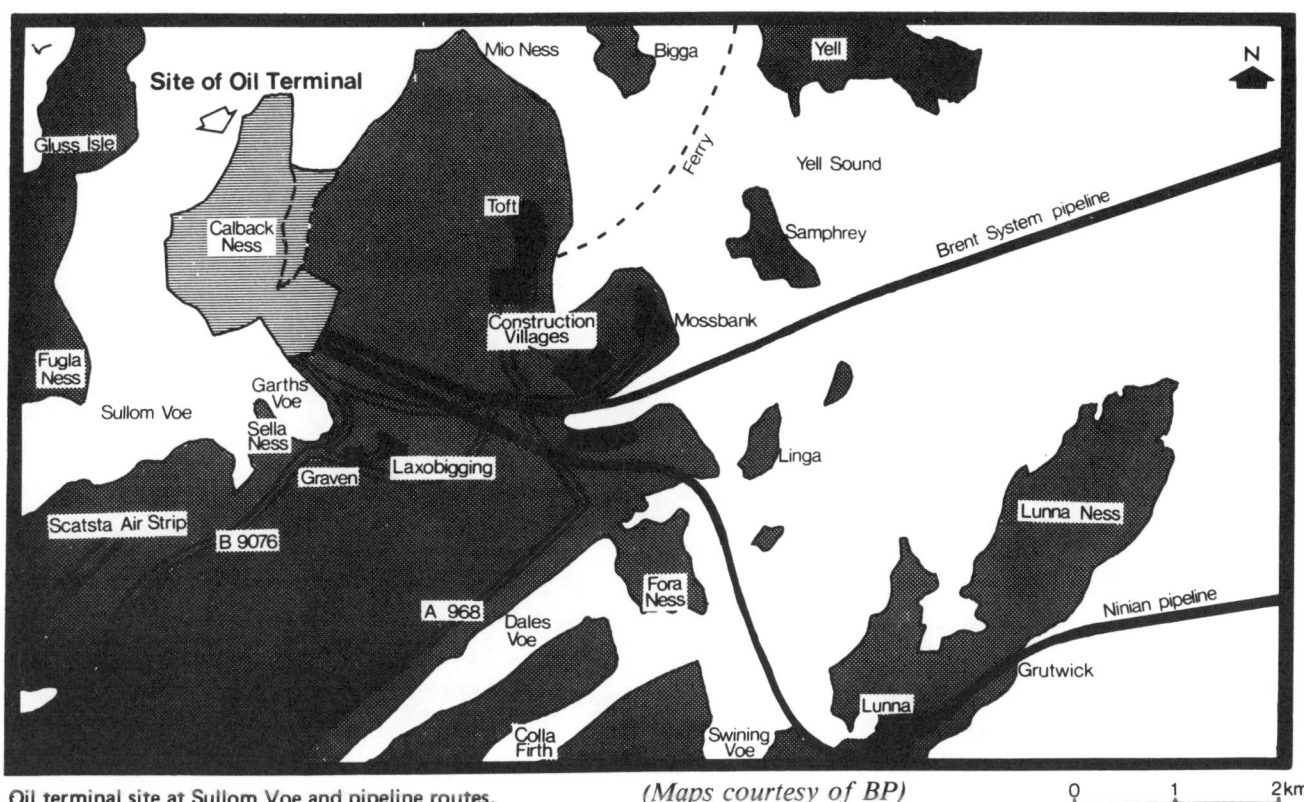

Oil terminal site at Sullom Voe and pipeline routes. *(Maps courtesy of BP)*

A. The Pipelines

The laying of the Brent and Ninian pipelines began in 1975 and was completed in 1976. The Brent and Ninian Pipelines come ashore at Firths Voe and Grutwick respectively, their paths converging a few hundred metres inland from Firths Voe (see map). Both pipelines were thereafter buried in a common trench at the side of the access road which leads to their destination within the Terminal.

(i) The Brent System

This consists of the Brent, Central Cormorant, Dunlin, Hutton, North Cormorant, North West Hutton, Murchison and Thistle oilfields being linked by feeder pipelines to the Cormorant A platform. (See map). On this platform there is the powerful pumping equipment necessary to give sufficient pressure to send the oil and associated gases from the Cormorant and above mentioned fields along a 36 inch (91cm) diameter pipeline, 96 miles to Sullom Voe. The first oil into Sullom Voe came in this pipeline from the Dunlin via the Cormorant Field when it became operational on 25th November 1978. The rest of the pipeline system is gradually being commissioned* as the various fields come on stream.** As the fields build up their production*** the pipeline throughput will increase towards its capacity of 1,050,000 barrels per day. From 1978 to Spring 1981 when commissioning of the process plant began at Sullom Voe, all the oil was degassed offshore, and came ashore in this "dead" state.

The 20 companies participating in this project through having shares in the oilfields concerned are —

Amerada Petroleum Corporation of the UK Limited

****Amoco UK Exploration Company

****British National Oil Corporation

BNOC (Alpha) Limited

Burmah Oil Exploration Limited

Charterhouse Petroleum Development Limited

****Conoco (UK) Limited

Deminex Oil and Gas (UK) Limited

Deminex (UK) Exploration and Production Limited

Esso Petroleum Company Limited

Gas Council (Exploration) Limited

Gulf Oil Corporation

Gulf (UK) Offshore Investment Limited

Mobil North Sea

North Sea Incorporated

Santa Fe Minerals (UK) Inc.

****Shell UK Limited

Tricentrol

Thistle Development Limited

Ultramar North Sea Limited

---

*See Table IV "Oil Pipeline Status Offshore Shetland as at December 1980".
**See Chart I, Chapter One, for further details.
***See Chart XVIII "The Brent and Ninian Pipeline Systems".
****Oilfield Licence Group Operators.

THE BRENT AND NINIAN PIPELINE SYSTEMS

Chart XVIII

Name of Field	Licence Block No.	Operator	Est. Recov. Reserves (m bbls)	Estimated Production Level '000 bbls				
				1981	1982	1983	1984	1985
1. Brent Pipeline System								
Brent	211/29	Shell	2000*	225	405	450	500	500
Cormorant	211/26 and	Shell	110 and 400	20	60	125	145	170
— South	211/21							
— North								
(Central areas also included in the reserves here)								
Dunlin	211/23 211/24	Shell	450-600	110	100	80	65	50
Hutton	211/28	Conoco	260	—	—	—	30	85
NW Hutton	211/27	Amoco	280	—	—	50	105	105
Murchison	211/19	Conoco	350-380	59	84	105	101	84
Thistle	211/18	BNOC	450-500	130	150	150	140	115
Sub Total (i)			4300-4530	544	799	960	1086*	1109*
2. Ninian Pipeline System**								
Heather	2/5	Union Oil	150	24	30	30	27	23
Ninian	3/3 3/8	Chevron	1200	310	335	325	260	230
Magnus	211/7a 211/12a	BP	450	—	—	25	90	129
Sub Total (ii)			1800	334	365	380	377	382
Total (i) + (ii)			6100-6330	878	1164	1340	1463	1491

(Production estimates based on Wood Mackenzie figures)

*Brent Field Reserves noted here include oil and associated gas liquids. Not all oil and associated gas produced by this field will necessarily come to Sullom Voe via the pipeline system. The Brent Field has the SPAR offshore loading buoy which is an alternative production facility and the FLAGS pipeline for gas and associated gas liquids to St. Fergus. (Pipeline capacity 1,050,000 barrels).

**It can be seen from these figures that with Fields presently producing into the Ninian Pipeline System, the system is far from its capacity of 950,000 barrels per day. It is probable that the new discovery at 3/7 and 3/8 near the Ninian Field (see Chart III — Unnamed Discoveries of Interest) will feed into and boost the Ninian System, and that possibly Alwyn Field, and other fields yet to be developed will too.

NOTES:
(i) 20,000 barrels per day is approximately 1 million tonnes per year.
(ii) m. bbls = million barrels.
(iii) Est. Recov. Reserves = estimated recoverable reserves — i.e. not all oil in place can be recovered; using present technology up to 50% of oil in place can be recovered from some of these fields — it depending on several factors eg. the geological structure, field pressures, type of oil, etc.

Brent Pipeline Costs

Owing to the fact that all the feeder lines from the other oilfields in the Brent System connect to the Cormorant A platform, this platform is partially owned by Shell/Exxon, the operator of the Cormorant Field, and partially by the other participants in the Brent pipeline who are named above. The cost of the pipeline from Cormorant Field to Sullom Voe was £1 million per mile.

The Construction of the Pipeline

(i) **Materials**

75,000 tonnes of 36 inch steel tubulars were obtained from Germany to build this pipeline. (A fair percentage of this pipeline was stored, handled and exported from Norscot Base, Lerwick — See Chapter Three). This represented considerably more material than was used in the Ninian pipeline, the reason for this being that there are more feeder pipelines in the main Brent system than there are in the Ninian System.

(ii) **Contractors**

The concrete covering and coal tar enamel for the pipeline was supplied by Bredero Price for the first 62 miles. J. Ray McDermott installed the concrete coated pipe using the laybarges LB23, LB28 and Choctaw II; Brown and Root's Barge BAR316 did the necessary trenching. The inshore seabed preparation was done by J. Williamson & Partners, and Land and Marine Engineering Limited, of Bromborough, were responsible for the landfall section of the pipeline. The landline runs for a total of five kilometres from its point of landing at Firths Voe, to Sullom Voe (see map), and was built in conjunction with Ninian landline by Press/Spie Batignolles.

The pumping plant for the Cormorant platform was provided by Mather & Platt, UK. Pipeline tie-ins have been completed as far as possible form the linking fields, but more tie-ins have yet to be completed.*

(ii) **The Ninian System**

This presently consists of pipeline connecting the Heather Field to the Ninian Field, and thence to Sullom Voe, with the tie-in from the template on Magnus Field to Ninian Field expected to be completed shortly.*

The first oil through this pipeline came from the Heather Field and arrived at Sullom Voe in December 1978. Pumps on the Ninian field are used to pump the oil, and associated gas (from 1981) to Sullom Voe from both Ninian and Heather Fields 105 miles (169 kilometres) to Sullom Voe along the 36 inch pipeline.

See Table IV "Oil Pipeline Status Offshore Shetland" in Chapter One.

There are 13 companies which are participating at present in the Ninian pipeline system through having interests in the fields concerned.* These are:—

The British National Oil Corporation

**BP Petroleum Development Limited

**Chevron Petroleum (UK) Limited

Det. Norske Oljeselskap A/S

Getty Oil (Britain) Limited

ICI Petroleum Limited

London & Scottish Marine Oil Company Limited

Murphy Petroleum Limited

Ocean Exploration Co. Limited

Ranger Oil (UK) Limited

Scottish Canadian Oil & Transportation Company Limited

Tenneco United Kingdom Inc.

**Unionoil Company of Great Britain

The Ninian Pipeline Costs

As with the Brent Pipeline, it is not possible to give an exact cost of the Ninian pipeline, as it has yet to be decided how many fields will be connected into this pipeline. All one can say is that subsea pipeline costs were approximately £1 million per mile. For tie-ins to other pipelines one needs additional pipe and risers, and often specially made connectors, and these are extremely expensive. It is assumed that other companies will participate in this system in the future; for example, Total, which may develop the Alwyn Field.

The Construction of the Pipeline

(i) **Materials**

60,000 tonnes of 36 inch steel tubular pipe for the 105 miles of pipeline to Sullom Voe was obtained from Japan.

(ii) **Contractors**

In the mid 1975 weather window, the laying of the Ninian pipeline began, using the 550 foot long semi-submersible lay barge, Viking Piper. Pipe was laid again in 1976, and the line was completed in record time, by 28th May 1976. All the pipe for this line passed through Lerwick. The pipeline was coated by M. K. Shand UK, and handled and stored at Norscot Base, Lerwick, prior to being taken offshore to the Viking Piper barge (Viking Offshore Pipeline SA of Switzerland were responsible for this part of the operation).

After the pipeline had been laid it was buried in the seabed and the bury barge Santa Fe Creek (owned by the Santa Fe Company) carried out this function. Land and Marine Engineering, completed the final phase of bringing the pipeline ashore. The Ninian Line makes two landfalls at Grutwick and at Firths Ness having re-entered the sea at Culness. As already mentioned, Press/Spie Batignolles completed the land line sections. Vickers Oceanics, using a two man mini submarine, were employed to travel the entire length of the line to check its condition before the line was pressurised for testing purposes, to ensure it was ready for the first flow of oil ashore in December 1978.

Field ownership shares can differ from pipeline shares.
**Oilfield Licence Group Operators.*

The Pipeline Operators

As previously noted, the pipeline operators share the same Terminal at Sullom Voe where the oil is received, processed, stored and exported. This means that each operator has a double responsibility — to take care of the interests of the group of companies participating in his own pipeline and to ensure the safe and efficient operation of this pipeline. This is done by —

(a) maintaining an operational interface between the platforms offshore and the Terminal;

(b) monitoring the reception and handling of the crude oil at the Terminal;

(c) liaison with Terminal operator on allocation of Terminal resources to participant companies.

Brent and Ninian Oil is kept in separate tanks but it could be co-mingled. The cost of common support facilities such as the Power Station and the administration of the Terminal are also shared. The cost of each pipeline and its gas processing and stabilisation facilities are paid for separately by each pipeline group.

The objective of the pipeline operators —

Shell for the Brent System, and BP for the Ninian System — is that both pipelines produce continuously without impeding production at the oilfields offshore.

OIL PRODUCTION THROUGH THE BRENT AND NINIAN PIPELINE SYSTEMS
*1978-1979-1980   Chart XIX

Pipeline System	Year	Cubic Metres	Barrels	**Metric Tonnes
Brent	1980	20,031,842	126,060,376	16,773,845
	1979	12,323,566	77,552,205	10,367,579
	1978	1,015,565	6,390,951	856,991
Ninian	1980	14,271,132	89,808,256	12,066,455
	1979	10,116,821	63,665,158	8,555,865
	1978	108,013	679,726	92,186
Brent + Ninian	1980	34,302,974	215,868,632	28,840,300
	1979	22,440,387	141,217,363	18,923,444
	1978	1,123,578	7,070,677	949,177

*From November only   **Please note units used

In order for the pipelines to transport the oil and associated gas there has to be a Terminal built to receive, store, process and export it.

The Construction of the Sullom Voe Terminal

The Oil Terminal includes the following main components:—

(i) Facilities for receiving and treating the incoming crude oil from the pipelines;

(ii) Oil and gas storage;

(iii) Facilities for receiving and treating oily ballast water from tankers;

(iv) Service facilities, including utilities such as the Power Station, control rooms, fire and safety support services, maintenance facilities, etc.

One of the first signs of the impending construction activity at Sullom Voe was the arrival of the surveyors Sir Alexander Gibb and Partners at Calback Ness in August 1974. To reach the site they could either drive in a Landrover (or similar vehicle) along the muddy shoreline at low tide or go across in a small open boat from Sella Ness (see map).

It was obvious much needed to be done before the construction project could get into full swing. Two main problems had to be solved:—

(i) The accommodation of a sizeable workforce at a remote location which had no facilities to cope with such an influx;

(ii) The transportation of materials, equipment and people to the construction site.

(I) Accommodation

The Zetland County Council had recognised that during the construction phase of the Brent and Ninian Pipeline Systems, and the Sullom Voe Terminal, there would be a need for accommodation for incoming workers, the local labour force being totally unable to meet the needs of such a large project. Figures given to the Council by the Oil Industry indicated that there would be a maximum requirement for the scope then envisaged for 1100 men to be housed in the Sullom Voe area. To meet this housing need, the Council decided to build a construction workers village, on a suitable area of land that it owned at Firth, near the main Sullom Voe Terminal site.

Firth Construction Village

To overcome the difficulties that the management of this village created the Zetland County Council formed a company, Grandmet (Shetland) Limited, in joint venture with Grandmet Scottish Site Services Limited, (a subsidiary company of the Grand Metropolitan Hotels Group), to operate the village. Three Directors were appointed by the Council and three from the Bateman Catering Organisation, being a division of the Grand Metropolitan Group. The Chairman of Grandmet (Shetland) Limited is one of the Council appointed Directors.

The Council contracted with Miller Construction Northern to build the construction village, named Firth Construction Village, and to provide certain fittings and furnishings.

The Council were insistent that the standard of accommodation should be high, especially since the men for whom it was designed were working on a bachelor status, far from home, and in a remote area which did not have the facilities to cope with such an influx of people. Every man was to have his own room with desk, wash basin and cupboards, and there were to be TV rooms and quiet rooms, bars, a shop, bank and medical centre, a large entertainment hall and fully equipped gymnasium and squash courts. There were also to be canteen and laundry facilities. The catering services and village management were to be provided by the Bateman Catering Organisation, through its subsidiary, Grandmet Scottish Site Services. All persons employed in the Village were therefore to be employees of Grandmet Scottish Site Services.

In July 1975, Grandmet (Shetland) Limited contracted with Shell UK, as Constructor of Sullom Voe Terminal, to provide accommodation and catering facilities for the incoming construction workers. This contract was later assigned to BP Trading, when BP assumed responsibility as Constructor of the Sullom Voe Terminal from Shell in December 1975. BP therefore, representing all the companies involved at the Terminal, is responsible for the Construction village contract.

To build Firth Construction Village in the first instance required an incoming labour force who had to be accommodated, and Miller Construction Northern (who built the village) established a pioneer camp at Firth in 1974. This pioneer camp accommodated not only Millers men but also other pioneers in the construction of the Sullom Voe Terminal — JMJ, Foster Wheeler, and Gibb. These men stayed at the pioneer camp until Firth Construction Village opened its doors on 4th August, 1975, its first residents being 130 men from JMJ.

Since August 1975, Firth has expanded to its present size, accommodating over 1100 men.

Grandmet (Shetland) Limited contributes an 'opportunity fee' to the Shetland Islands Council, and these payments together with a share of the profits of Grandmet (Shetland) Limited are paid into the Shetland Islands Council Charitable Trust.

Toft Construction Village

Hardly had the initial buildings at Firth Construction Village been erected when, at the end of 1975, the Oil Industry informed the Council, through the Sullom Voe Association, that they had revised their projected labour requirements in line with the increase in the scope of the Terminal and thought to have a peak labour force of 2200 men by early 1978. This would entail the provision of a second Construction Village as the first (Firth) would be full by the end of 1976.

Aerial view of Firth Construction Village with Toft Construction Village in background, looking north in October 1978.
(Photograph courtesy of BP)

In addition to the 2200 men for the actual construction work, there would be 450 professional and supervisory staff and 200 resident catering personnel.

To cope with these numbers, a total of approximately 2900 rooms would be required. As the existing construction village at Firth had an accommodation facility to cope with a maximum of approximately 1100 men a second construction village with an 1800 room capacity was planned.*

Toft was chosen as the site of this village in order to provide the least disturbance to the local community. For example, traffic from Toft Site to Calback Ness would not pass through any villages or groups of houses (see map).

The land on which Toft stands was acquired by the Shetland Islands Council, and leased back to the Industry as Constructor of the Terminal. BP is in charge of the organisation and development of this village.

Like Firth Village, Toft is operated like a hotel in the sense that rooms are allocated for the period of a man's stay, his gear is put into store at the end of his tour of duty, and a new room allocated on his return from leave. In this way maximum use of the accommodation has been achieved.

A fundamental point about both villages is that the major part of the construction is erected on a temporary basis. An undertaking has been given to the Shetland Islands Council that Toft will be dismantled at the end of the construction phase of the Sullom Voe Terminal and Port with the exception of any buildings that the Council may ask to be left.

A special form of construction that does not entail removing the peat and thus disfiguring the landscape has been used for some of the buildings at Toft. This involves the lighter buildings in the village (the accommodation modules) being placed on piled foundations.

The construction of the villages was undertaken in phases. At Toft, each phase constituted blocks of 580 bedrooms with their associated facilities such as snack bars, and recreation centre including gymnasium and halls. The first occupation occurred in June 1977, and the village was completed in 1978. Miller Construction Northern, the company which built Firth Village, was also contracted to build Toft. It was responsible for the entire complex. To enable rapid building, Terrapin prefabricated as much of the structure as was practicable, and this material was imported across Toft pier to minimise the impact on the local roads.

The total project was estimated to cost about £10 million (1977 values). The managing contractor in the village is Tarmac National Limited, a subsidiary of Tarmac. Tarmac have contracted Taylorplan to provide catering, cleaning and laundry services.

Unlike the Firth Village project, Shetland Islands Council did not undertake any responsibility for the construction or management of Toft.

The growth of the workforce in both Firth and Toft Construction Villages led to an increased need for leisure and recreational activities, especially entertainment. The buildings had been provided for these activities, and organisation to arrange and operate them to residents requirements, needed further development. In early 1977, therefore, a group of Firth residents got together and decided that they would need to be actively involved if these services were to meet their needs. This resulted in the formation of the Sullom Voe Social Club in April 1977, drawing its membership and committee from the workforce. The Club has a turnover in excess of £2 million per year, stemming from the operation of the bars at Firth and Toft, and the Toft shop. Overheads such as wages and operating costs, account for a substantial proportion of this £2 million but enough is left over to provide sports and entertainment every night. Cabaret, bingo, films, darts and snooker leagues, dances, Christmas parties for local children, and other activities are run by the Club in addition to a very active sporting section including angling, rugby and football teams and boxing. BP assist with expenses for new equipment and certain running costs.

The Construction Villages have other Committees in operation concerned with general aspects of communal living, on which each company with residents in the villages is represented, as are BP and the Social Club in order that any complaints can be dealt with on a regular basis, and living conditions are made as harmonious as possible. There are also Chaplins to the workforce with church services and active pastoral work is evident.

By February 1981, Firth had 1167 men, Toft 2237, both Construction Villages having expanded on the original accommodation estimates.

The Accommodation Ships

The saga of temporary accommodation was not to be at an end, with the completion of the two construction villages. During 1978 the scope of work had again increased as the capacity of the Terminal was revised upwards from 826,000 barrels per day to 1,390,000 barrels per day. This meant additional processing plant and tanks and the Industry advised the Shetland Islands Council that the time-space for construction was tight and additional workers were required to handle the peak requirements of the job. The labour force at Sullom Voe had already swelled to 4,800 by this time (80 per cent working on site at any one time — work being on a rota basis to allow for periodic leave), and the oil industry indicated that they required an extra 700 people to bring total workforce numbers on the Island to 5,500. The solution was sought in the provision of two accommodation ships, and the Council agreed to grant a works licence for the first ship, the 350 berth Rangatira, to be moored in Garths Voe, adjacent to the Terminal.

The Rangatira had previously worked as an accommodation ship at Kishorn, where the central concrete platform for Chevron's Ninian Field was constructed by Howard Doris. A visit had been made in early 1976 by representatives from the Shetland Islands Council and the Technical Working Group of the Sullom Voe Association to see how this ship was used. It had been proposed to use an accommodation ship at that time but Toft Construction Village was instead given priority. The Council, whilst agreeing no planning permission was needed for accommodation ships (only works licences) were in principle not really in favour of them, but having been given assurances as to the standards which would be operated, agreed to temporarily grant temporary works licences to permit the ships to be moored.*

The Rangatira arrived in Shetland in September 1978, and was for use by engineers and contractors staff. The ship had been specially converted in Glasgow docks so that each cabin was for a single occupant (some occupants cabins were also retained for them when they were on leave so they could return to the same cabin) and to match the high standards in the existing villages. Part of the car deck was converted for use as a sports room and keep fit centre, and self contained sewerage and laundry systems were added.

Following the satisfactory experience of operating Rangatira, the Council agreed to the second ship, the Stena Baltica, which arrived in March 1979, and left in January 1981. During that period the ships provided in total accommodation for 700 professional and technical staff involved in the Terminal Construction. Grandmet (Shetland) Limited have provided accommodation services on both ships. It is popularly said in Shetland that these have provided two of the best restaurants in the islands.

The ships had (and Rangatira still has) their own Social Club for the residents, providing films and entertainment. The sports facilities at the two Construction Villages were (and are) also open to the ships residents, as they are for all Sullom Voe workers.

Other Accommodation

Other accommodation for senior staff involved in the Sullom Voe Project has been provided by BP and the Shetland Islands Council. The Council built houses at Brae, Mossbank, Firth and Voe, originally intended for the long term people involved in the operation of the Terminal. Due to delays in the completion of the Terminal, these houses were available before the operations staff arrived in Shetland. The Council therefore decided to let the houses at an economic rent on a short lease basis (up to two years) to BP, and not to individuals, BP would then allocate the houses as they deemed fit. (See Chapter on "Infrastructure" for further details). BP also built themselves 50 temporary chalets at Upper Lea, Firth for use by construction and commissioning personnel, in addition to over 100 permanent houses.

(II) Transportation of Materials and Equipment for the Sullom Voe Terminal Project

Two critical items were required to ensure ease of transportation of materials, equipment, and people, to build the Sullom Voe Terminal:— an access road to the site; and a construction jetty to bring in the very heavy loads — of up to 500 tons in weight or 120 feet in length — which were expected. Heavy loads were expected, not only for items such as power station boilers — but also because inherent in the organisation of the construction at Sullom Voe was the principle that, where possible, prefabrication would be carried out on the UK mainland, rather than construction on the site. No normal roads could be expected to cope with such traffic thus in 1975 a Construction Jetty was built.

It was in its capacity as Harbour Authority at Sullom Voe that the Shetland Islands Council granted the works licences for the accommodation ships.

The two accommodation ships moored in Garths Voe. Stena Baltica (left) and Rangatira in April 1979.

(Photograph courtesy of BP)

Arrival of first Pre-Assembled Unit (PAU) at Sullom Voe Construction Jetty. This PAU contains heat-exchange equipment for the crude oil stabilisation process. This PAU is approximately 12 metres in each direction and weighs 330 tonnes.

The Construction Jetty

The Construction Jetty, situated on the North East Shore of Sullom Voe, was built by George Dew's and JMJ in 1975. A design for rapid construction was chosen and the jetty, consisting of rock filled circular cell type cofferdams (a honeycomb type of concept) was joined to the land by a rubble causeway. All this was surmounted by a concrete surface and road base which made a concrete ramp (invaluable for roll on roll off cargoes), an integral part of the jetty.

This conventional ro-ro ramp was altered in 1978 to accommodate the Pre Assembled Units (PAU's) and Pre Assembled Piperacks (PAR's), being exceptionally heavy (up to 480 tonnes) and awkward loads to discharge. Otherwise the jetty which has an area of 328 by 67 ft, was to remain unaltered throughout what was to become a most impressive logistics operation.

A tremendous operation began, in which, across the relatively small jetty, over a million tonnes of cargo were handled in the four and a half years from July 1976 to the end of 1980* using a peak dock labour force of 38 men during the busiest period in 1978 and 1979.

Bases were established on the mainland to control shipping from the ports of origin — which were mainly Leith, Grangemouth, Bromborough and Invergordon. Invergordon was used almost exclusively for shipping sand and aggregates. The other ports shipped more general cargoes.

Only 49 ships — or three per cent of all these handled — were diverted to other ports because the Construction jetty was too busy. At the height of operations the jetty was manned 24 hours a day and a constant VHF radio watch kept. In this way, a thousand ships on a thousand days (each ship not usually exceeding 3,000 tonnes), were handled by this relatively small jetty.

See Chart XX "Construction Jetty Traffic 1976-1980". Also Chart XXI "Total Tonnages over the Construction Jetty for Contractors and Others involved in the Sullom Voe Terminal Project (July 1976 to December 1980").

SULLOM VOE CONSTRUCTION JETTY OPERATIONS (July 1976 — December 1980)

Chart XX

Year	Bulk Sand	Vessels	Bulk Cement	Vessels	Bulk Aggregates	Vessels	Armour Stone	Vessels	P.A.U.'s P.A.R.'s	Vessels	General Cargo Inward	General Cargo Outward	Vessels	Total Tonnes	Total Vessels
1976	10185	14	1793	7	14347	11	1434	3	NIL	—	13633	3297	70	44689	105
1977	59836	37	16059	36	81088	47	7624	16	NIL	—	70831	6856	212	242294	348
1978	142702	68	43451	97	76094	35	520	1	1676	4	91616	9007	225	365066	430
1979	128908	75	26218	46	NIL	—	NIL	—	6630	20	68384	15397	195	245537	336
1980	59850	59	8007	24	NIL	—	NIL	—	8353	14	33645	14666	115	124531	212
Totals	401491	253	95528	210	171529	93	9578	20	16659	38	278109*	49223	817	1022117	1431

NOTES:

*This includes all steel needed for the tankage, power station and other major facilities on site.

Vessels Diverted: 49

Jetty: 328' x 67'

Labour: 1976 1 Foreman 11 Dockers
       1977 1 Foreman 11 Dockers (Jan/Mar)
                 1 Foreman 14 Dockers (Mar/Jul)
                 1 Foreman 17 Dockers (Jul/Dec)
       1978 2 Foremen 36 Dockers
       1979 2 Foremen 36 Dockers
       1980 2 Foremen 20 Dockers

(Source of information: Wimpey Marine Limited)

TOTAL TONNAGES MOVED OVER THE CONSTRUCTION JETTY FOR CONTRACTORS AND OTHERS INVOLVED IN THE SULLOM VOE TERMINAL PROJECT (July 1976 to December 1980)

Total Number of Consignees: 123 Chart XXI

Contractor	Tonnes	Contractor	Tonnes
Foster Wheeler	168,622	Val de Travers	118
J.M.J.	7,723	Trentham	32
Tarmac	38,019	Hunter Construction	38
Christiani & Neilsen	33,437	Moore, Barrett & Redwood	35
BP Petroleum Development Ltd.	6,876	Marjon Plant	14
J. D. Tighe	14,583	J. Nadin	389
Boyton	510	Deborah Scaffolding	8
Matthew Hall	374	Sunters/ITM	667
Wimpey M. E. & C.	1,021	Sullom Voe Engineering	130
Knockbreda	1,618	Marinco	55
C.B. Scaffolding	4,601	Capper Neill	3,197
Numire	39	Harvey Plant	451
Shetland Line	1,088	Thistle Engineering (LWK)	1
Condor (Scotland)	3,682	E.P.L.	32
Pressure Test Services	202	Morceau F.P.	1,003
P&O Ferries	725	Lilley (LWK)	12
S. & J. D. Robertson	20	R. B. Farquhar	6
Keith Bros	7	G. Biro	15
Grangemouth Heavy Lifting	29	Streeters	65
N.G. Bailey	557	ARC	287
Macbon	24	A.Q.S.	81,991
Terrapin	5	Quality Inspection	7
Shetland Towage	12	Lindsay Shipping	2
Flaregas	24	Rubery-Owen	19
G. Mundy	6	L.J.K.	530,553
Donald Dunbar	12	Motherwell Bridge	34,857
Mercer	276	Millers	7,232
Northern Divers	32	Wimpey Marine	467
P. Fraenkel	4	George Dew	3,974
Rossedin Buildings	1	Rotary Services	223
Air Products	7	Holliday Hall	1,027
Universal Rock Drilling	10	C.J.B.	50,569
Kinross Plant	331	Wm. Press	7,877
D.J.E. Seeding	839	BP Oil Pollution Equipment	2,008
G.E.C.	63	Shetland Islands Council	112
Ruberoid	312	S.G.B. Scaffolding	889
Econ. Forestry Services	23	Cape Contracts	1,832
DECCA Radar	7	Hudsons	69
Precision Insulation	52	Boyd & Shearer	2

Contractor	Tonnes	Contractor	Tonnes
Austin Hall	934	Dawson-Keith	422
Crosshill Engineering	33	Binns Fencing	8
Sparrows	412	Old Rob	4
Scottish Land Plant Hire	2	Northern Lighthouse Board	15
Tillon Travers	4	Shell UK	49
Thomas Storey	63	Mundie Plant	12
G.P.O.	11	Sutherland Transport	14
Sullom Voe Social Club	670	Encore Drilling	1
Abird Superior	32	Cool Heat	1
Jonvald Marine	1	Soil Mechanics	19
Laurie & Millar	5	E.F.G. Newlands	30
McLean Quinn Plant	69	Tate Pipe Linings	87
Edarn Plant	26	Capt. Turner c/o S.I.C.	3
R.M.B. Electric	2	Loyne (Belfast)	52
Glenlec Ltd.	1	British Sailors Society	51
Freeman-Morrison	288	Sub-Work Diving	20
Sullom Quarries	32	Cementation (LWK)	3
Rigging International	1,483	J. Finnis	1
Barrett	1	Ozalid	2
Albion Fencing	101	Greenham	202
Kelvin Catering	654	Davy Plant	42
Grandmet	37	Rangatira	112
Colt International	20		

Total Tonnes: 1,022,071

The movement of the PAU's and PAR's and some of the other unusual cargoes required special techniques. I.T.M. Sunters and Rigging International were contracted to transport these cargoes. Millipede like trailers, some with as many as 190 wheels, each wheel of which could be independently or collectively controlled, were used to offload these units from the barges on which they had been shipped to Sullom Voe. The barge also required to be carefully ballasted whilst the unloading proceeded.

Although the stevedoring labour was entirely locally employed, Wimpey International provided accountants and Wimpey Marine Limited of Great Yarmouth provided other necessary office management. The most noticeable feature of this entire construction jetty operation has been that comparatively few people — at the peak activity in 1979, 56 were employed by Wimpey — had provided this service.

The Transportation of people to the Sullom Voe Terminal Construction Site

BP, as Constructors of the Terminal, contracted with Foster Wheeler (GB) Ltd. to provide transportation services to and from the Construction site. At first people came to Sumburgh Airport, 55 miles and two hours by bus from the Terminal, by scheduled flights. Later, charters were increasingly used. By 1977 the problem had become acute. Sir Frederick Snow and Partners were commissioned to study whether the former Royal Air Force airstrip at Scatsta could be revived and used as an airport for Sullom Voe workers. This eventually occurred in 1978 (see Chapter Four — Scatsta Airport — for details).

Foster Wheeler (GB) also provide a workers bus service which transports people from outlying areas to and from their work at the Terminal.

Peat, 1976. *(Photograph courtesy of BP)*

The Management of the Construction of the Sullom Voe Terminal

Having discussed the accommodation and logistics services, which were necessary before construction at the Terminal could begin, it is to the organisation and management of this £1200 million project that this book now turns.

The Terminal is being constructed on a site extending over a 1000 acres to handle a throughput of 1.41 million barrels per day of crude oil.* Whilst it is true that there has been no radically new technology employed at this Terminal and that all the plant is of well tried designs developed by the Oil Industry, the sheer scale of the project has generated managerial and organisational challenges. This is particularly so if one considers that since 1978, the Construction of the Terminal has continued in parallel with Operations, and with Commissioning activities.

One therefore has three activities, involving three separate groups of people, present on the site at the same time.

A Sullom Voe Project Organisation has been evolved by BP Petroleum Development as Constructor of the Terminal (see Chart). BP's Construction Manager acts as Chairman of the Contractors Committee which meets frequently to make sure that construction requirements are adequately co-ordinated.

Most of the civil engineering work, which involved the excavation of vast quantities of peat and rocky subsoil for the pipeline, oil terminal and industrial area developments, was completed by 1978, by which time five million cubic metres of peat and six million cubic metres of subsoil had been moved. This civil engineering work has been done by JMJ, who were joined in mid 1976 by Lilley and Keir, to form the LJK consortium.

Construction of the Terminal

It was planned to bring ashore 'dead' or de gassed crude oil as soon as practicable and load it by gravity from crude storage tanks into ships, to permit an early startup of production through the Terminal.

See Layout of Terminal Diagram.

SULLOM VOE CONSTRUCTION PROJECT ORGANISATION (as at 1980)

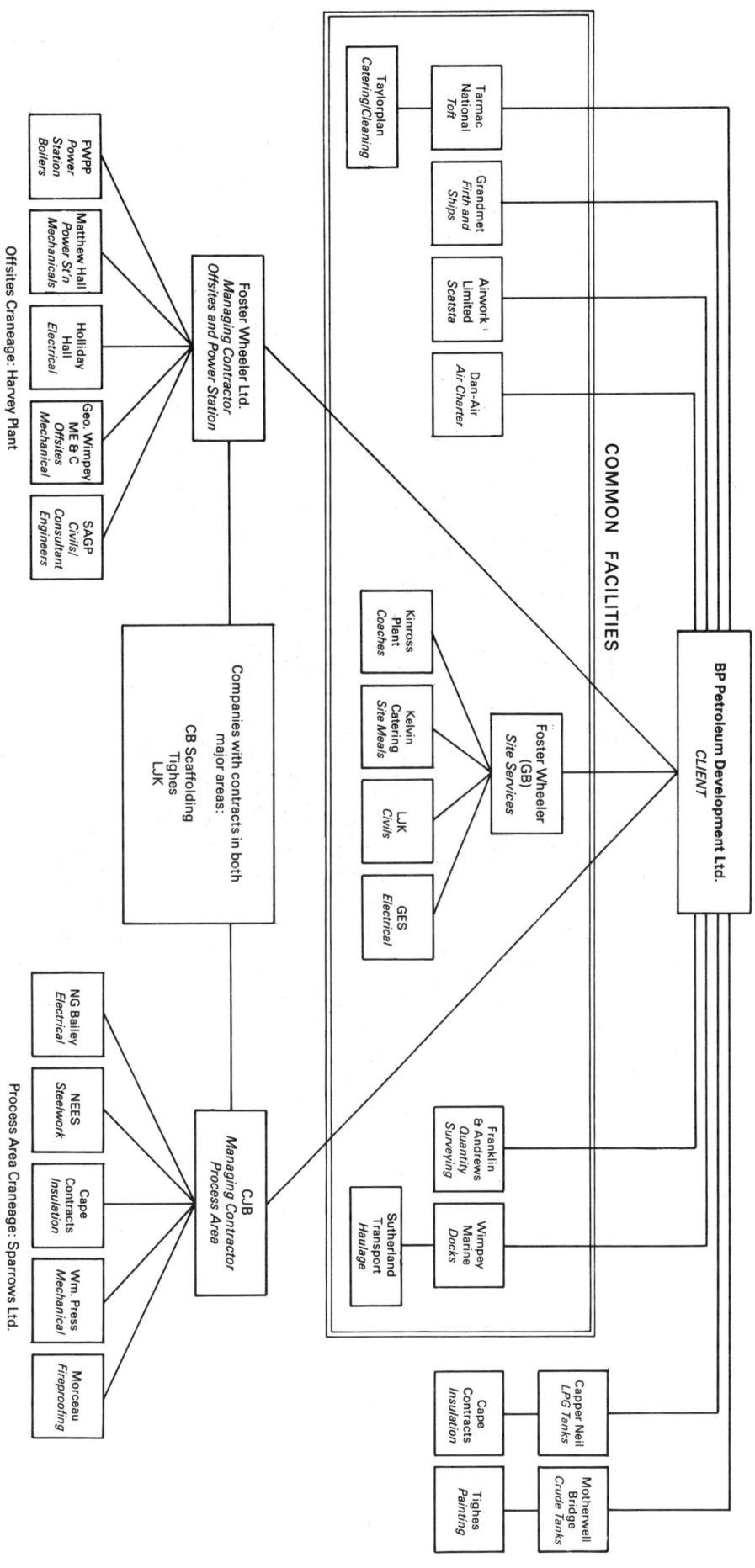

Offsites Craneage: Harvey Plant

Process Area Craneage: Sparrows Ltd.

Building the £1200 million Oil Terminal to a tight schedule has meant spending over £600,000 every working day, and employing numerous large contractors, and many smaller ones, to ensure that progress matches expenditure.

The Organisation chart shows the reporting relationships of the main employers on the site, but, for the sake of simplicity, does not include many of the smaller companies and contractors who played an important part in the development of Sullom Voe.

The Chart shows BP as Constructor and Client, but it must be remembered that they act for the whole consortium of thirty-three oil companies with varying financial interests in the offshore fields that supply and are paying for the Terminal.

Contracts at site have been either directly with BP as in the case of Motherwell Bridge and Dan-Air, or through one of the two managing contractors, Foster Wheeler Ltd. or CJB. Foster Wheeler have responsibility for the offsites and power station, which includes the LJK earth moving and civil engineering contract; Wimpey Mechanicals work on pipelines around the site, and Holliday Hall in both areas, as electrical and instrument contractors. CJB have had responsibility for the Process Area where oil and gas will be separated, and the LPG Chilldown Area, where gases are cooled for storage. These areas employ LJK on civil engineering work, William Press on mechanical, and Bailey's on electrical and instrumentation contracts. In addition Foster Wheeler (GB) Ltd. provide the common user site services, administering Kelvin's site catering, the Kinross Plant coaches, and the maintenance of construction buildings across the site.

115

LAYOUT OF THE SULLOM VOE OIL TERMINAL

SULLOM VOE OIL TERMINAL

Situation — Calback Ness, 31 miles north of Lerwick (by road).

Area — Approximately 1000 acres.

Construction Phase — February 1975 to 1982 (for presently planned facilities).

FACILITIES PRESENTLY ENVISAGED AND UNDER CONSTRUCTION

1. **Power Station**

    Initially 125 megawatt capacity.   Steam Boilers: Total Number: 5
    2 x 75 tonne/hour units
    3 x 110 tonne/hour waste heat boilers

2. **Process Facilities**

    1 surge tank for Brent pipeline
    1 surge tank for Ninian pipeline   each capacity 200,000 barrels

    5 stabilisation trains: each rated at 330,000 barrels per day

    6 high pressure compression trains   3 refrigeration and fraction trains
    5 low pressure compression trains   1 spare refirgeration train

    4 gas and liquid drying trains

3. **LPG Facilities**

    3 LPG chilldown trains
    5 fixed roof storage tanks: each capacity of 30,000 cubic metres of 20,000 tonnes.

4. **Surface Crude Storage Tanks**

    Total number: 16

    Tank dimensions — Capacity: 100,000 cubic metres gross, approximately 90,000 cubic metres working; Diameter: 76 metres; Height: 22 metres (with floating roof).

    Each tank has a capacity of 80,000 tonnes oil = 600,000 bbls = 21 million Imperial gallons.
    Capacity of 16 tanks = 9.6 million bbls = 336 million gallons.

    There is one bund for every two tanks. Volume of each bund = 110,000 cubic metres.

5. **Ballast Water Storage Tanks**

    Total number: 4. Capacity: 66,000 cubic metres; Diameter: 76 metres; Height: 22 metres + 1 metre wind skirt.

6. **Flare System**

    One 105 metre multiple flare stack on Hill of Garth (see diagram); minor flares at surge tanks and LPG storage area.

7. **Effluent Treatment Plant**

    Capable of treating an effluent throughput of 5000 cubic metres/hour.

    Facilities include — two interceptors, two recovered oil tanks, five Sand Filters, three emulsion break tanks.

    Average content of oil in water at outfall = 10.15 PPM.

8. **Throughput of Terminal**

    More than 800,000 barrels per day at early 1981. (Present terminal is designed to handle about 1.41 million barrels per day: it could be expanded to handle up to 3 million barrels per day if a third pipeline comes ashore in Shetland.

JETTIES

1. **Construction Jetty**

    To handle materials for construction, in ships up to a maximum of 3,000 tonnes dwt.

2. **Crude Oil Jetties**

    Total present number: 4

    No. 1 Jetty: Tankers from 18,000 to 12,000 tons dwt.*   **No. 3 Jetty:** Tankers from 30,000 to 300,000 tons dwt.
    No. 2 Jetty: Tankers from 50,000 to 300,000 tons dwt.   **No. 4 Jetty:** Tankers from 30,000 to 300,000 tons dwt.

Oil and Liquefied Petroleum Gas.

This plan dictated to some extent the order in which construction proceeded at Sullom Voe:—

1. **Crude Storage and Ballast Water Tanks**

    The provision of two ballast water and four crude oil storage tanks (as outlined in the Overall Development Plan) was a priority, and in Autumn 1975, Motherwell Bridge Engineering moved onto the site to begin constructing the tanks. Since 1976, when the concept of overground storage was agreed, Motherwell Bridge have won additional contracts, building in total 16 crude oil tanks, four ballast water tanks, two surge tanks (literally to take any "surges" or irregular influxes from the pipelines) and 14 others for various functions in the power station and effluent treatment areas. In 1978, ten of these smaller tanks ranging in size up to 15 metres high, 17 metres in diameter, and weighing 90 tonnes were shipped over the construction jetty as complete prefabricated units and erected on the site.

2. **The Power Station**

    The Power Station building was started by Tarmac in September 1976. It has a total generating capacity of 125 megawatts, and provides electrical power for the entire site.

    The Power Station was planned so that it could use either gas or diesel as fuel, although it is intended mainly to fuel it with gas separated from the Crude Oil coming into the Terminal. Two package boilers to generate steam and five twenty-five megawatt gas turbine generators, the exhaust gases from which will also be used to raise steam in waste heat boilers built by Foster Wheeler Power Products, who manufactured the two gas/diesel fired boilers; these and the five gas turbines can all be fuelled with either gas or diesel. The Power Station building is the largest building in Shetland, and has nearly 19 miles of pipe work provided by Matthew-Hall, to carry gas and water in and around it, in addition to steam and condensate to and from the process and effluent treatment areas. Electrical power is used, amongst other things, to pump oil to the storage tanks and also for the loading pumps to the jetties.

    Outwith the power station and process plant areas, Holliday Hall were contracted to provide electrical and instrumentation requirements. The work began in September 1977 and is expected to be complete in 1981. Power is distributed around the site by 33 thousand volt (33kV) cables to three substations where it is transformed to 11kV and then to 22 other substations where it can be reduced further as required. The complete system comprises over 75 kilometres of electrical cables and 1100 kilometres of instrument cables, reflecting the fact that this is one of the major installations in Britain.

3. **The Effluent Treatment Area**

    This was another area for priority completion, when operations began at the Terminal in 1978, although the ballast water treatment facilities were not in fact ready until 1979.

    The Effluent Treatment Area includes ballast water tanks for draining and cleaning oily ballast from the tankers which come in to load at the jetties.

    Beneath the surface of the Terminal LJK have built drainage systems to carry both clean rain water into the sea and contaminated water from parts of the site such as the process area. The contaminated water is processed with the oily ballast water through interceptors, the water eventually passing into a holding basin thence to the sea, and the oil and oily water to emulsion break tanks and sand filters, from which the oil is recovered.

4. **The Process Area**

    This was the last of the major areas of the Terminal to be started. Civil engineering work to prepare the level of the site was begun by LJK in 1977. Drainage systems and the first foundations followed to enable the first steel erection in 1978. The majority of the piperacks and process equipment were prefabricated (in module form) on the UK mainland and shipped in over the construction jetty, with only a little steelwork and the large compressor houses erected on the site. Thirty-six Pre Assembled Units (PAU's) containing heat exchangers or vessels plus all their associated pipe work valves, insulation and instrumentation were fabricated at yards such as Redpath Dorman Long on Teesside, and Charlton Leslie at Wallsend.

    Seventeen additional large items of equipment, mainly gas compressors, were also packaged in steelwork for shipping to Shetland. Thirty-six Pre Assembled Piperacks (PAR's) complete with piping, insulation, etc., were contracted to companies such as Cleveland Bridge Engineering on Teesside and William Press Production Systems on Tyneside. The PAR's were finally lifted into position by a 1000 ton capacity Rosencranz crane, one of the world's largest.

General view of process area and power station after two weeks rain, October 1976. *(Photograph courtesy of BP)*

Aerial view looking north-east towards the power station with the process area behind, September 1980. *(Photograph courtesy of BP)*

5. **The LPG Tank Area**

Construction of the Liquid Petroleum Gas (LPG) tanks was begun in Autumn 1977 by Capper Neill. They used inflatable air domes, which had a giant tent like appearance, to cover the tank bases whilst insulation blocks of styro-foam were laid. The gas in the tanks will be liquid at down to minus 50 degrees Centigrade, therefore insulation is essential to prevent either the ground freezing or the liquid warming up. The tank pads are also heated to plus 1 degree Centigrade to prevent ground deformation stressing the steel of the tanks. By the middle of 1979 Cape Insulation began the work of enclosing these tanks in galvanised steel cladding with insulation material foamed between the two layers. In front of the tanks William Press are carrying out mechanical work in the LPG chilldown area where the pure LPG's from the process area are cooled before storage.

There are of course other areas of note within the Terminal, such as the Administrative Centre, the Control Centres, and the Fire Station, but the major construction of the Terminal (excluding the jetties which comprise part of the port) has been concentrated on the above noted facilities. Two of the other areas of note within the Terminal merit particular attention, the fire and safety, and the control systems. These can be considered together, since terminal fire and safety monitoring is tied into computers in the control rooms, of which there are two.

The first is the pipeline and processing control room (P.P.C.R.), located in a blast proof building outside the process area: the functions of this are two-fold, to maintain contact with offshore production platforms and to liaise on the flow of oil to Sullom Voe Terminal, and to control the operation of the process area. The P.P.C.R. monitors the meters which ensure the integrity of the pipelines and measure the arriving crude, and follow and route the oil and gas through the various processes which are described elsewhere. The second control room, also in a blast proof building, controls the flow of oil and gas to and from the storage tanks and to the jetties, the loading of tankers, and the operation of the effluent treatment area.

Both control rooms monitor a wide range of sophisticated equipment, including meters, pumps, valves, etc., and a great many heat, smoke and gas detectors, located all round the Terminal. These detectors are duplicated in an emergency services control centre located at the fire station and, in the event of an incident, it is from here that fire and medical support is co-ordinated. All the control rooms are in close contact at all times ensuring the security and safety of the Terminal and all those employed in it. A fire water ring main provides fire vehicles with a high pressure water supply anywhere on the Terminal and this main is charged by a pump house on Jetty 2, with a duplicate on Jetty 4.

Millions of manhours have been required to construct the Sullom Voe Terminal, including the land pipelines and Harbour works, in this, Europe's largest construction project.

The Operation of the Sullom Voe Terminal

As stated, Terminal operations officially began with the first oil ashore in November 25th 1978. Full facilities were not available and a basic Terminal operation was mounted with the de-gassed or dead crude from the Brent and Ninian System pipelines being transported at the natural pressure from the pumps offshore straight into the crude oil storage tanks. These tanks have a head of 30 metres above the ships into which the oil was then loaded by gravity feed.*

There were no de-ballasting facilities available and ships arrived and left with their ballast aboard.

The Donovania, which was the first tanker in (see Chapter Six) did practice runs to help achieve a terminal/port interface. By the end of the first year's operations, in November 1979, 380,000 barrels per day was coming through the system with six crude oil storage tanks and two jetties in use, and power being provided from the Power Station was supplemented by temporary electricity generation around the Terminal. At about the same time ballast water treatment facilities came into operation, commissioning having been in progress for about two months beforehand. From November 1979, therefore, all ships coming into Sullom Voe could discharge their ballast water, which is often oily from previous crude oil cargoes having been carried in the same tanks.

**It is not unknown for Oil Terminals to use this system of gravity loading tankage. For example, it is quite common in the Middle East.*

Aerial view of flare stack looking south-east across the process area, February 1981.

(Photograph courtesy of BP)

BP Operations Firecrew and their appliances, 1978. *(Photograph courtesy of BP)*

The oily water is pumped by the ship to the ballast water storage tanks, where it is allowed to settle over a period of up to 48 hours. Generally a good separation can be affected in 24 hours. Oil floating on the water is skimmed off, put into "recovered oil" tanks and injected into the crude being put on the ships.

The water left is put into oil/water interceptors*, and after some residence time and additional skimming is then pumped through sand filters where any oil remaining will be absorbed onto the sand grains on the sand filter beds. The water is then put into a final holding basin before being pumped offshore into Yell Sound, through an effluent diffuser.

The resulting effluent has, with the level of operations at the Terminal to date, contained oil of an average concentration of six parts per million parts of water. A monitoring programme, developed from biological baseline studies,** is in operation, with regular monitoring being carried out by Heriot Watt University, which reports to the Shetland Oil Terminal Environmental Advisory Group (SOTEAG).*** Major monitoring studies are carried out twice per year at the present time, and more often if there are oil spills.

---

*Interceptors are large horizontal troughs with baffles, weirs and skimmers, where most of the oil content separates out.
**Baseline ecological and environmental studies began in 1974/75 and were published in May 1976 in the report "Oil Terminal at Sullom Voe — Environmental Impact Assessment" by the Sullom Voe Environmental Advisory Group (SVEAG).
***See Sullom Voe Association Organisational Chart; also Appendix "Other Organisations connected with the Sullom Voe Oil Terminal".

The final stage in the development of Operations at the Sullom Voe Terminal will come when the gas treatment facilities are commissioned and it is possible for the Terminal to accept 'live crude' oil (ie. crude oil with gas dissolved in it) through pipelines.

Oil, gas and water are three normal components found in the oil fields offshore. When oil is being produced, much of the water is removed at the platform, but it is not always possible or indeed desirable to remove all the gas. The gas which is not required for platform fuel* is dissolved under pressure in the crude oil which is then delivered to the Terminal through the pipelines. The crude oil first arrives at the 'pig' receiving station in the Terminal. The 'pig' receiving station is where pipeline inspection gadgets or 'pigs', which are mechanical scrapers or pipe cleaners running along the pipeline with the oil flow, are 'received' and taken out.

Crude oil can leave waxy deposits on the walls of the pipe and this is removed by the 'pigs'. The oil then flows to the pipeline integrity meters where volume, temperature and pressure are measured and compared with the offshore readings. If there is a leak in the pipeline system, there will be a discrepancy in the readings, and the line must be checked to determine the cause.

The oil and gas mixture arrives under pressure. If for any reason there is a surge in this pressure that could damage downstream equipment in the process area, or if there is a failure in the process area so that oil cannot be handled safely, then automatic valves divert the flow to the surge drums and tanks which stand over the process area. Here the gas, which would normally be separated and processed further, is simply removed and led off to either the main flare or the surge flares where it can be safely burnt. Each tank has a capacity of 200,000 barrels, equivalent to four hours maximum flow from one pipeline, allowing sufficient time for an orderly shut-down of the system.

Stabilisation

The incoming crude oil is first heated in three stages, using stabilised crude oil, condensed steam and live steam. The hot oil passes to the dehydrators where water is removed and the hot, dry oil then flows to the separator vessels. The pressure of the oil and gas mixture is reduced in two stages causing the gas to separate from the oil. The stabilised (gas free) oil is then in a safe condition for storage in the crude oil tanks or for loading to tankers and so undergoes no further treatment.

Compression

The separated gas is a mixture of methane and ethane (often called 'natural' gas), propane and butane ('Calor' gas or LPG). The LPGs are more valuable as pure products than as fuel for the power station and have to be separated from the other gases. To do this, the original mixture has to be at high pressure and low temperature.

From the separators, the gas passes to the compression area where it is compressed in three stages. The compressed gases are then measured, cooled and dried to remove any last traces of water before they pass to the fractionation area.

Fractionation

In the fractionation area, the mixture of gases is distilled in a series of three 140 ft. high columns. The first is the de-ethaniser where methane and ethane are separated, appearing as gas from the top of the column while the propane and butane appear as liquid at the bottom. The methane/ethane mixture is led off to the power station where it is used as fuel gas for the gas turbines.

The second column, the de-propaniser, produces propane gas at its top. Similarly, butane is produced in the de-butaniser, where any heavier compounds (pentane and higher hydrocarbons) are removed as liquid which is added back to the stabilised crude oil. The pure propane and butane are then treated in separate units to remove any sulphur compounds that may be present, metered, and then flow to the LPG chilldown area where they are liquefied and stored before being shipped out in special gas tankers.

First fractionation will take place early in 1981 and will include three stabilisation trains, three compression trains, one fractionation train, one chilldown train and four LPG tanks.

Some may be removed to fuel the gas turbines on the platforms for power generation.

Aerial view of Calback Ness before construction of Sullom Voe Terminal, 1974. *(Photograph courtesy of BP)*

Aerial view of construction of Sullom Voe Terminal on Calback Ness, looking north-west, February 1981. *(Photograph courtesy of BP)*

EMPLOYMENT AT SULLOM VOE TERMINAL 1978-1980

Chart XXII

Comparison of Forecasts and Actuals — (i) Terminal Construction — 1978/79/80

Quarter	1978 1st	2nd	3rd	4th	1979 1st	2nd	3rd	4th	1980 1st	2nd	3rd	4th
CATERING												
Forecast	797	880	1073	973	1095	1243	1194	1317	1326	1277	1301	1233
Actual	812	998	1005	1117	1211	1341	1329	1308	1282	1368	1285	
OIL INDUSTRY CONTRACTORS												
Forecast	2688	3151	3663	3805	4002	4288	5332	4862	4853	5024	*5341	5051
Actual	3049	3519	3866	3949	4256	5297	5315	5254	5343	5669	5816	
TOTALS												
Forecast	3485	4031	4736	4778	5097	5531	6526	6179	6179	6301	6642	6284
Actual	3861	4517	4871	5066	5467	6638	6644	6562	6625	7037	7101	
Error	+376	+486	+135	+288	+370	+1107	+118	+383	+446	+736	+459	

NB: Contractors include S.I.C. Contractors.

*Amended.

Source of figures: Joint Employment Monitoring Group.

(ii) Terminal Operations
Long Term Employment Opportunities at Sullom Voe Terminal

Situation at February 1981

No. of long term jobs: 773

No. of locally employed people: approximately 50% overall.

Future increase in job opportunities: limited by the rate of natural wastage.

Sullom Voe Company/Organisation	No. of Employees
BP Petroleum Development Limited	377
Sullom Voe Engineering*	
Moore, Barrett and Redwood**	296
Commercial Catering***	

*Sullom Voe Engineering Limited is part of the Wood Group and has a maintenance contract at Sullom Voe with BP.

**Moore, Barrett and Redwood work in the Terminal laboratory and assay the crude oil and other substances connected with the operation of the Terminal.

***Commercial Catering supply catering services for all operations staff.

Chapter Six

The Port of Sullom Voe

The Shetland Islands Council is the Harbour Authority of the Port of Sullom Voe under the terms of the Zetland County Council Act 1974. Apart from the statutory duties as Harbour Authority, the Shetland Islands Council has obligations under the Port and Harbour Agreement made with the Oil Industry, and signed on 15th March, 1978. Some of the main obligations under this Agreement are as follows:

The Shetland Islands Council

(i) are required to complete the construction of the four tanker loading jetties and to appropriate these for the exclusive use of the oil companies;

(ii) may be required to construct additional jetties and appropriate them for the preferential use of the companies (i.e. the Council may nominate users of these jetties when they are not required by the Companies);

(iii) must permit the Companies to allow other pipeline group members to use appropriate oil jetties;

(iv) are required to appropriate the Construction jetty to the Companies;

(v) must maintain the Harbour, its facilities, the approach channel, and the jetties;

(vi) must seek to finance the construction of the jetties;

(vii) subject to any voluntary or statutory rights of relief, must meet the cost of dealing with oil pollution for which the companies are not responsible under the agreement;*

(viii) must secure the provision of a mooring boat service.

Apart from obligations incumbent on the Shetland Islands Council, the other party to the agreement, in this case the Oil Companies operating into Sullom Voe, also have obligations. The oil industry are obliged to reimburse the authority for the provision and appropriation of the four jetties.

This reimbursement consists of the Council receiving each year the costs of providing the jetties plus

(a) 2 percent profit on the "Actual Capital Cost", i.e. the cost of securing the construction of the jetties, and

(b) 1p per long ton (2240 lbs) of crude oil and liquefied petroleum gas exported over the oil jetties.

These payments are protected against inflation by complex indexation provisions, based on the international crude oil prices, and adjusted for any quality changes in the oil exported. If such indexation proves impracticable, adjustments are made in line with the Retail Prices Index.

It is perhaps impracticable to make a precise estimate of the profit payments over the life of the Agreement, but it has been estimated by various sources, that between 1978 and the year 2000, payments could total at least £48 million.

The oil companies have indemnified local fishermen against the possibility of damage to fishing grounds arising from the Terminal effluent discharge into Yell Sound.

The Agreement runs initially until 31st August 2000. The companies have options to extend it, only in respect of the oil jetties, but not beyond 31st August 2050. The companies have an option to discontinue any number of oil jetties at 31st August 2000 only. If this option is exercised, the Agreement cannot be extended beyond August 2025.

Having briefly stated the respective obligations of the Council and the Oil Industry, it is now appropriate to note the more practical side of the day to day operation of the port. As Harbour and Pilotage Authority, the Shetland Islands Council through their marine staff at Sullom Voe regulate the movement of tankers within the Pilotage district and the harbour area (see map). The Port handles all sizes of tankers from 20,000 to 350,000 tons deadweight, the largest to date being of 311,000 tons deadweight.

The Shetland Islands Council is owner of the land at Sullom Voe upon which the Terminal is constructed. In return for a lease of the land the Council will receive an appropriate rental. The structures for which the Shetland Islands Council are directly responsible, as a service to the shipping industry are four tanker jetties, a tug jetty, and small craft harbour, a dry cargo jetty with ramp (the "construction jetty"), and the Port Administration buildings at Sella Ness, and the navigational aids and services.* The tanker loading jetties, built with the aid of a loan from the European Investment Bank, will eventually enable four large tankers to be loading oil or LPG** at the same time. Loading facilities for liquid petroleum gas are expected to be available from Autumn 1981, on one jetty, and there are processing facilities for receiving oil polluted ballast water from tankers at all four jetties. The loading facilities and all systems in the transfer of liquids and gases installed on the jetties are provided by, and are the responsibility of, the Oil Industry.

The Operation of the Port of Sullom Voe Pilotage Services

From 1st February, 1978, a compulsory Pilotage district was created whereupon it became the statutory duty of the Shetland Islands Council as the Pilotage Authority, to provide a pilotage service within the Sullom Voe District (see Tanker Routes map, showing pilotage limits). This in conjunction with the Council's powers as Harbour Authority gives Shetland Islands Council effective control of shipping within these waters for the purposes of safe navigation. Certain classes of vessel are exempt from compulsory pilotage, within the Sullom Voe Shetland Pilotage District, which includes HM Ships and other vessels, which are exempt under the conditions laid down in the Sullom Voe Pilotage Byelaws.***

The Sullom Voe Pilotage District includes the waters of Yell Sound and Sullom Voe, encompassing all creeks, voes and inlets (see map).

Qualified Marine Officers, whose duties include acts of pilotage, are available on a 24 hour basis at Port Control and maintain a continuous radar and VHF radio watch on all vessels using the Port.

The movement of tugs, pilot boats and other craft are monitored from the Port Control room — which is equipped with six channel VHF radio and high definition surveillance radar.

There are two pilot boarding areas located in Yell Sound. Large vessels rendezvous with the pilot at the Northern entrance to Yell Sound, near the Point of Fethaland (see map), about 12 miles from the tanker jetties. Small vessels can use the South East Channel and rendezvous with the pilot off the entrance to Tofts Voe (see map).

The Authority has two pilot craft for use at the Port of Sullom Voe and a third boat is under construction.

Pilot Boats	Overall Length	Service Speed
(1) Sullom Mareel	53 ft	16 knots
(2) Sullom Spindrift	65 ft	20 knots

It has been estimated that these facilities have cost over £50 million to provide.
**Only one tanker can load LPG at a time.*
***Masters of British registered vessels of less than 1600 Gross Registered Tons may be eligible to sit an examination to gain a Pilotage Certificate under the conditions laid down in the Sullom Voe Pilotage Byelaws.*

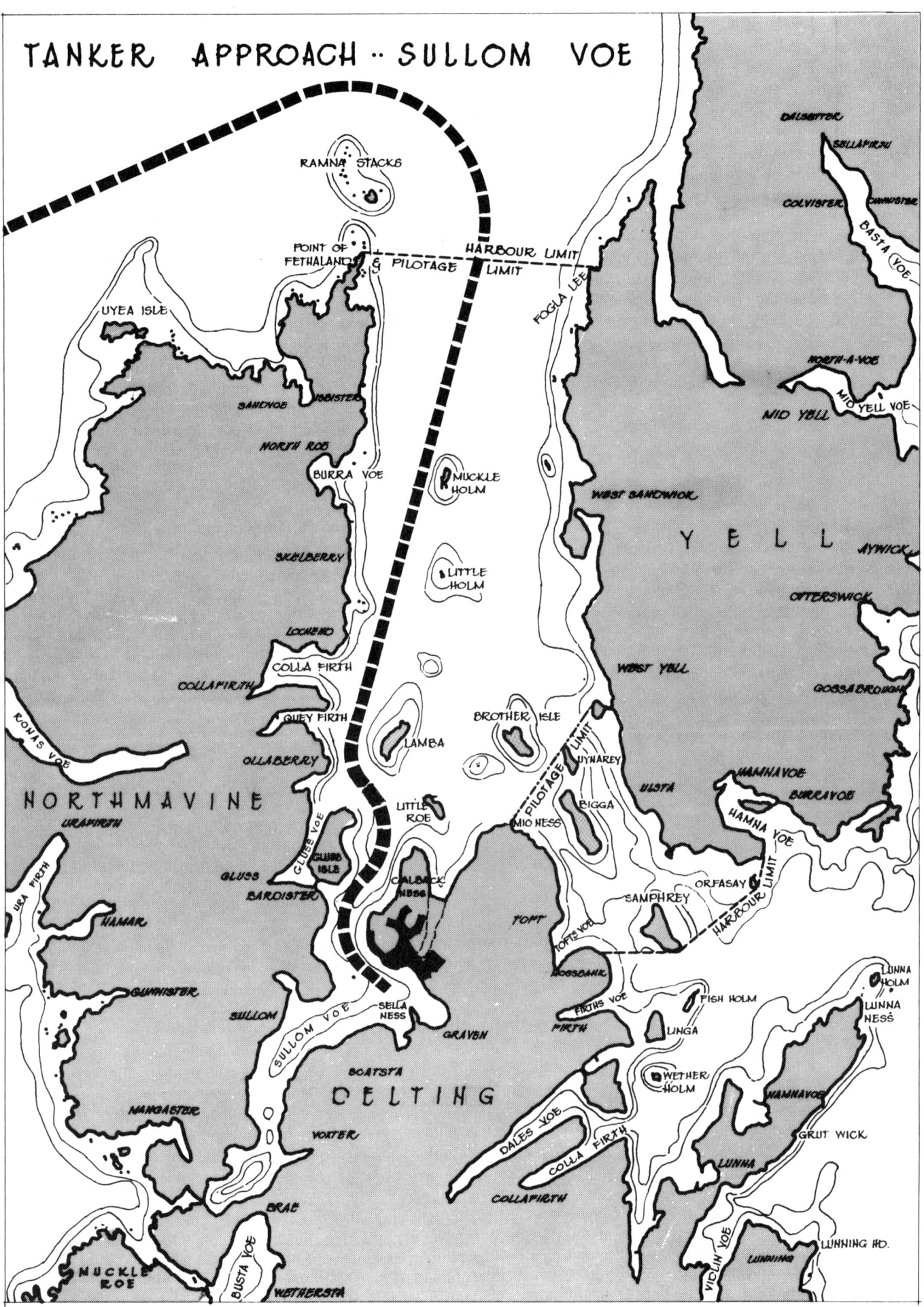

Both boats are of fibreglass construction and are fully equipped with modern navigational aids. They are also equipped with life saving appliances and oil pollution dispersal spray gear.

Pilotage Authority Requirements

Vessels bound for Sullom Voe are required under the operational procedure to provide the Pilotage Authority with the following information:

(i) An estimated time of arrival at least 24 hours prior to arrival or leaving the last port — whichever is later,

(ii) Confirmation of the vessel's estimated time of arrival at least six hours before arrival at the seaward limit of the Pilotage District and thereafter any significant changes in excess of one hour.

(iii) On coming within range of Sullom Voe Harbour Radio, the estimated time of arrival to be given by VHF radio.*

Vessels arriving from the south are requested to advise Sullom Voe Port Control of the time of passing latitude 58 degrees 20 minutes North, indicating their exact positions and intended courses and speed until arrival at the Pilot Station. Ships approaching from any other direction are requested to provide similar information upon reaching a position 200 miles from the Pilot Station. Prior to sailing from Sullom Voe, all vessels are requested by the Port Authority to indicate intended outward bound courses until south of latitude 58 degrees 20 minutes north or 200 miles from the Pilot Station, as appropriate.

Shetland Islands Council and the Oil Industry have also agreed that all tankers proceeding to or sailing from Sullom Voe should keep at least ten miles off the Shetland Islands coastline. As Fair Isle** is part of the Shetland Islands, it is recommended that vessels navigate the Fair Isle channels approximately midway between Fair Isle and either Mainland Shetland or North Ronaldsay, Orkney. Additionally, it is recommended that passage between off lying islands or skerries, including between Foula and Mainland Shetland (on the West side of Shetland) should not be attempted. Navigational recommendations also give boundaries outwith which tankers should steer when either approaching or leaving Yell Sound.

If for any reason of safety or emergency the tanker Master cannot comply with the foregoing recommendations it is requested to immediately notify his intentions, with reasons, to Sullom Voe Port Control by VHF Channel 16 or via agents. Charts are also not to be cleaned off until the tanker has berthed at Sullom Voe in order that the Port Authority may check the route the tanker has followed.

The above measures and recommendations in respect of routes to be taken when approaching the Pilotage District of Sullom Voe form part of a comprehensive range of precautions that the Shetland Islands Council has devised since tankers began operating into the port, for the export of crude oil at the end of November, 1978.

As stated in the previous chapter, ballast water treatment facilities are available for tankers but these were not operational until a year after the commencement of crude exporting operations, in November 1979. Within several months after tanker loading operations began at the end of November 1978, polluted beaches and damaged wildlife were reported around the shores of Orkney and Shetland, particularly on the west coastline. This suggested that certain vessels were disregarding the IMCO*** conventions on Oil Pollution and discharging dirty ballast water and tank washings, whilst approaching or immediately after sailing from Sullom Voe.

Such practices could not be accepted by the UK Department of Trade, the Shetland Islands Council**** nor the Terminal Operator. These practices were also against the established internationally agreed rules. In order to reduce the risks of oil pollution from these sources, aerial surveillance***** was introduced in March 1979, in the waters around Shetland and Orkney and stringent rules governing the disposal of oil polluted ballast were

---

*Channel 19, call sign Sullom Voe Terminal is used for initial contact with ships.

**Fair Isle is situated 25 miles south of Sumburgh Head (see map of Shetland, Chapter One).

***Rules are agreed through the Intergovernmental Maritime Consultative Organisation, whose Headquarters are based in London.

****As local authority, the Shetland Islands Council has certain statutory obligations to clean up pollution.

*****Air Ecosse provided the aerial surveillance service from 14:3:79 to 16:8:79, Peregrine Air Services from 21:8:79 to 16:11:79 and London Air Taxis from 20:11:79 to the present time.

devised. Briefly these rules, again agreed between all the parties involved, forbid the discharge of any ballast contained in the cargo spaces of tankers, except to the shore facilities at Sullom Voe. Tankers must also arrive at Sullom Voe with at least 35 percent of their total cargo carrrying capacity, in the form of ballast. Such disciplines, combined with aerial surveillance, have practically eliminated such incidents.

There is no doubt that the provision of the Terminal's ballast water treatment facilities in late 1979, destroyed any further incentive for vessels to discharge dirty ballast at sea. There are certain consequences for those tanker masters who disregard or abuse these anti-pollution measures. Any vessel sighted discharging dirty ballast, oily waste or other pollutants, will be immediately reported to the UK Government for appropriate action to be taken. In addition, locally at the Port there may be extensive delays, before services and berths are available at Sullom Voe, whilst investigations by the Shetland Islands Council and the Oil Industry are undertaken. The oil pollution surveillance aircraft carries sensitive photographic equipment which can produce high definition pictures in cases where vessels are suspected of disregarding the rules and there is always a Pollution Control Officer of the Shetland Islands Council, on board the aircraft to ensure accurate, responsible results. The surveillance flights take place as often as is required, and there is no scheduled flight pattern. If a suspect vessel in known to be arriving, or investigations of incidents are underway, the plane may be airborne at frequent intervals. In addition, the helicopter crews flying continuously to the East Shetland Basin, and other aircraft around Shetland have proved helpful to the Authorities, in reporting any possible oil slicks or ships contravening the agreed procedures.

The Shetland Islands Council is also very much aware of the dangers to navigation around the Shetland coast, due largely to outdated charts, offshore obstructions, strong tides, rapid weather changes, and the presence of local and international fishing vessels, perhaps with their nets spread over a wide area. Navigation within ten miles of the coastline as previously stated, has therefore been vigorously discouraged, and a Marine Circular was issued in November 1979 to emphasise this recommendation.

The Shetland Islands being somewhat remote, and sparsely populated in world wide terms, might be expected to accept greater risks from the pollution activities of certain tankers, than other areas. However, this is not the case, and it would be difficult for the tankers using Sullom Voe to go unobserved if they failed to comply with tanker routeing recommendations around the islands. Those tankers who deliberately pollute the waters around the Islands have found this out to their cost, especially when the repercussions of delayed entry to the Port or cancellation of loading have been applied, whilst investigations proceed. Sightings of irregularities have been made from the outlying Islands of Foula, Papa Stour, the Muckle Flugga lighthouse, Out Skerries, and Fair Isle, and action taken both by the Shetland Islands Council and the Oil Industry as a result.

Minimum Requirements for Tankers

Apart from routeing and navigational requirements, the Shetland Islands Council, as Harbour Authority require that all vessels which have been programmed for Sullom Voe should meet certain minimum standards. These standards generally comply with the recommendations of the Council of the European Committee for Vessels Transiting Community Seaports. Vessels should be fitted with:

(i) Fully operational radar;

(ii) Fully operational VHF radio with Channels 16, 14, 20, 12, 10, and 9;

(iii) Sufficient crew to handle tugs' lines and moorings efficiently;

(iv) Efficient propulsion and deck machinery;

(v) Efficient and adequate moorings to secure vessels safety;

(vi) Efficient signalling equipment;

(vii) Fully operational Tachometer, Rudder and Helm indicators;

(viii) Efficient mooring stations communications systems;

(ix) An efficient anchor windlass, and anchors;

(x) A complete and valid set of Safety Certificates;

(xi) All Officers to be properly qualified and carry recognised certificates of competency.

If any vessel fails to have these minimum requirements, delays can result. The Pilot boarding the Tanker will normally check that these requirements are met. If he has any reason to doubt this, the ship's entry to the Port will be delayed, and the Department of Trade inspector,* who has been resident in Shetland since January 1980, can be called in to inspect the ship and aid in confirming or otherwise its ability to comply with the regulations.

Finally, before arrival, all vessels intending to berth at Sullom Voe are required to provide Sullom Voe Harbour Radio with the following information:

(i) Name of vessel;

(ii) Port of Registry;

(iii) Name of Master;

(iv) Gross registered tonnage;

(v) Vessels length overall;

(vi) Draught fore and aft;

(vii) Whether loaded or in ballast; if loaded, nature of cargo and quantity for discharge;

(viii) Whether vessel is carrying segregated ballast;

(ix) Location where segregated ballast was loaded;

(x) Segregated ballast quantity;

(xi) Dirty ballast quantity;

(xii) Which Pilot Station (i.e. Northern or Southern entrances to Yell Sound — all tankers will use the Northern approach);

(xiii) Pilot boarding arrangements;

(xiv) Number of crew;

(xv) Last port of call.

Parts of this information will be used for varying purposes. The Shetland Islands Council has been particularly anxious that tankers calling at Sullom Voe are well found, well managed and crewed ships, there having been a few instances when ships chartered to call at Sullom Voe have been found to be lacking in those areas and in the minimum requirements.

To assist the Authority in preparing for the arrival of tankers and to improve safety measures a computer telelink service has been installed at Port Control. The Mardata** system was put in operation in May 1980. This system allows its users to key into a central data bank by telephone link and obtain information about tankers, including their operational track record.

Mardata consists of a consolidated list of the history of the vessels and uses Lloyds Intelligence Service as its basis. Accident damage, equipment bases, ownership and name changes are some of the details available. Whilst Mardata is not totally comprehensive, either in the inventory of accidents or vessels own record, it is one of the best information systems of its kind presently available. Those ships which have a poor record, for example those which have been involved in frequent incidents or incurred regular damage, are investigated.

In its fight against oil pollution the Shetland Islands Council has had the co-operation of the Oil Industry in the

---

*The Department of Trade inspector is a UK Government appointee who ensures that ships are complying with various statutes in respect of oil pollution, overloading, fire fighting and life saving.

**Mardata is the trade name of a commercial Company which operates a marine intelligence service.

measures it has taken. An example of this co-operation is where tanker masters who breach the ten mile zone can be subject to commercial sanctions. If it is found there is no genuine reason for a breach of this limit or there is a refusal to answer, and the ship's management or condition is suspect, the ship will be fully investigated before it is considered safe for it to enter the Port.

Further, in respect of pollution in the waters around Shetland there is incorporated in the charter party and FOB sale agreement, conditions which enable the oil companies to reserve the right to refuse cargo to the ship and to delay loading until certain steps have been taken or, if appropriate to cancel the charter, in certain circumstances.

Presuming that fair weather prevails,* and all the Port's requirements have been met, the vessel will rendezvous with the Pilot boat and pick up the pilot at the Northern end of Yell Sound near Grunay Isle. If the sea state warrants it, and provided the ship has the necessary facilities, it is possible the pilot may board the ship by helicopter. During the ensuing 12 miles run down Yell Sound and into the Sullom Voe Harbour, the vessel's speed is progressively reduced to below five knots. Tugs will join the vessel off Lamba Isle (see map) and, using their own lines, will secure themselves to the vessel with the aid of the ship's crew. The pilot will be at all times in radio contact with the tugs. After the bow and stern tugs have been made fast to the ship, it will reduce its speed, and then the two alongside tugs will be secured. The tankers are then turned around in the "turning area" and the mooring boats take their lines to the mooring dolphins, so that the tankers are berthed facing the direction from which they have just come.

After mooring, the ship will be boarded by the Terminal's loading master and the loading arms will be connected up by jetty operators, who are also employed by the Terminal, who are responsible for the loading operation. A ship's agent will then board, to take care of documentation, and possibly HM Customs, and others.

Vessels are obliged to produce their log book to confirm the amount of ballast they were carrying when they left their port of origin, with quantities of segregated ballast and cargo tank ballast indicated separately. All segregated ballast tanks are inspected by Shetland Islands Council and Oil Industry staff to see if they are clean. The Shetland Islands Council, as Harbour Authority, also takes random samples. Analysis is made of these samples to ascertain the quantity of hydrocarbon, coliforms, and heavy metal pollutants within the ballast. The Harbour Authority may, as a result of information so obtained, refuse permission for the discharge into the Harbour of segregated ballast water taken aboard in certain locations. The discharge of ballast water is not permitted unless with the consent of the Harbour Master. If pollution occurs, from any vessel within the Harbour Limits, the Oil Industry is obliged to provide the equipment and clean up the oil spilt.

Pollution clean-up equipment provided by the Oil Industry and stationed at Sullom Voe, is available to cope with oil spills of up to 2000 tonnes.** Beyond that, outside aid would be called upon. A series of procedures have been agreed by the Council and the Oil Industry, should pollution or emergencies occur, there being three publications (for internal use) in which procedures are laid down:

(a) Sullom Voe Harbour Authority — Emergency Plan.

(b) Shetland Islands Council — Anti-Oil Pollution Scheme.

(c) Sullom Voe Harbour — Oil Spill Plan.

The Management of the Port

There are a number of UK Ports, apart from Sullom Voe, where the local authority also acts as a Harbour Authority and is responsible for the operations of a major port — for example, this situation also pertains to the port of Bristol.

Shetland Islands Council established a Ports and Harbours Committee in 1975 and it is through this Committee that matters concerning the Port of Sullom Voe are channelled. The Director of Ports and Harbours, who is also

---

The Harbour Master operates certain criteria before the Port of Sullom Voe is closed for the berthing of tankers and gas carriers when wind speeds are consistently gusting in excess of 30 knots.

**See Appendix for list of equipment to cope with oil pollution.*

318,805 dwt Shell tanker 'Litiopa' being brought alongside crude oil loading jetty No. 2 by tugs of Shetand Towage, July 1980. *(Photograph courtesy of BP)*

the Harbourmaster of the Port of Sullom Voe and Pilot Master for the Sullom Voe Pilotage District, reports to the Ports and Harbours Committee, whose recommendations are thereafter ratified by the full Council.

Approximately forty percent of the 200 or so people directly employed in the operation of the Port (this number of 200 discounts BP's tanker jetty operatives) are employees of the Shetland Islands Council Department of Ports and Harbours, under the management of the Director, who reports to the Chief Executive of the Shetland Islands Council.

There are four other Committees of importance which the Director of Ports and Harbours Department is a member of — Zetland Harbour Advisory Committee, which was established by the Zetland County Council Act 1974; the Pilotage Committee; the Harbour Users Committee and the Sullom Voe Oil Spills Advisory Committee (see Appendix for further details).

The Zetland Harbour Advisory Committee was established by the Secretary of State for Scotland, under the terms of the ZCC Act 1974* and was constituted in 1980. The Sullom Voe Oil Spills Advisory Committee is a joint Shetland Islands Council/Oil Industry Committee which has a remit to recommend equipment and procedures to help counteract pollution.

Having discussed the Port and its operation, it is appropriate to consider some of the companies who are active in providing a service to the vessels using it.

Shetland Towage Ltd.

The fleet of tugs which provide a towage service for vessels at the Port of Sullom Voe are stationed at the Sella Ness jetty adjacent to the Port Administration buildings. This fleet is operated by Shetland Towage Ltd.

Shetland Towage Ltd. is a towage company formed by a joint venture between Shetland Islands Council and two shipping companies — the Clyde Shipping Co., and Cory Ship Towage (Clyde) Ltd. The company was incorporated on 14th February, 1975, negotiations for its establishment having begun in September 1974. The company has eight Board members — four of them from the Council, and four from the other companies involved, the Chairman being selected from one of the four Council directors. The object of the Company is to provide towage at the port and they do this at present by means of four twin screw tugs. It has now been agreed that there will be five tugs — the three which were purpose built at Hall Russell's shipyard in Aberdeen for use in the Port in 1978, and two which will be purchased. These will have propulsion units of the water tractor type.

The Stanechakker, Swaabie and Lyrie are twin screw 3800 horsepower vessels, named after Shetland birds, each having a bollard pull of 55 tons and a speed of 14 knots. They are fully equipped as fire fighting tugs and pollution control vessels. At present, seconded to this permanent fleet is a fourth single screw tug, the Flying Childers, with a bollard pull of 38.6 tons for towing off the hook and a 14.2 knot speed but without a fire fighting capability.

In 1980 Shetland Towage Ltd. had over 80 employees, including 12 tug crews of six men, plus four extra men to allow for training, and used to man the tugs. Of these, over 80 per cent are Shetlanders, the non-Shetlanders being found amongst the Masters and the Engineers. There are also shore based personnel for radio communications, stores and clerical work.

BP Petroleum Development Ltd., as Terminal Operator appointed to act on behalf of the Brent and Ninian Pipeline System Group, was deeply involved in the negotiations which led up to the provision of tugs by the Towage company. As Terminal Operator BP is concerned with operational aspects of the Industry's use of the tugs which are owned by the Towage Company. This factor, amongst others, led to protracted negotiations before all parties were satisfied, and the Heads of Agreement for the operation of the towage service were signed in July 1976.

See Appendix for further details.

SULLOM VOE PORT STATISTICS

Chart XXIII

Year	IMPORTS			EXPORTS			Total Throughput (i) + (ii)	Comments
	Coastwise	Foreign	(i) Sub Total	Coastwise	Foreign	(ii) Sub Total		
1980	206,528.52	4,916.96	211,445.48	12,007,585.99	16,305,044.59	28,312,630.58	28,524,076.06	99.21% (28,298,463.59) tonnes of exports were crude petroleum. 51% increase (9,620,139.59 tonnes) in amount of crude petroleum exported since 1979; 57.62% of crude petroleum exports were foreign going.
1979	274,183.62	4787.92	278,971.54	6,345,950.37	13,355,545.41	19,701,485.78	19,980,467.34	93.48% (18,678,324 tonnes) of exports were of crude petroleum; 67% of crude petroleum exports were foreign going.
1978	337,941.66	1337.931	338,279.59	540,412.01	246,687.67	787,099.68	1,125,379.271	Crude petroleum exports only began on 28th November, hence low figures compared to other years. 99.16% (780,518.87 tonnes) of exports were of crude petroleum. 31.61% of crude petroleum exports were foreign going.

135

The Shipping Agents

There are several ships agents offering services at the Port of Sullom Voe.

Ships agents are responsible for paying all expenses a ship incurs whilst in port, including Harbour Dues. They represent the shipowners or charterers, take care of customs documentation, crew changes, ships stores and generally offer a 24 hour, seven day per week service.

Ships agents are normally "put in funds" by the owner or charterer of the ship, and paid before the ship they are handling sets sail. This is important, if one considers that the dues for harbour services for a 300,000 ton foreign going tanker could be £80,000. Similarly, ships, which require investigation or repair or are in some manner delayed in port, can build up substantial dues for port services. The dues are levied for four day periods. Smaller ships, and the ships travelling to other UK ports pay less — for example, a 35,000 Gross Registered Tons ship would probably pay about £20,000 for services during a short stay (no longer than four days).

Agents fees are on the scale agreed by the Institute of Chartered Shipbrokers, which is based on the deadweight tonnage of the ship handled. There are variations within this, in that terms are negotiable with the major oil companies if they undertake long term contracts with agents for their ships rather than one off assignments.

The agency business at the Port is expanding as the number of ships calling increases. For example, with the LPG ships coming into Sullom Voe, traffic is expected to increase by about 25 percent.

Ships Agents

The largest of the shipping agents at Sullom Voe is Hay & Co. Ltd., who employ ten people at their office in the Port Administration Buildings. Since the Port opened in November 1978, they have handled over 600 tankers, and have the agency for Shell, Esso, BP, Gulf and Chevron ships, and have handled many others besides.

The business is competitive and in 1980, two other companies arrived on the scene — Roxburgh Henderson & Co. who are part of the Ocean Transport and Trading Group; and Escombe McGrath, who are allied to the P&O Ferries Group. There are other agents in the Port, but they tend not to handle tankers to any marked extent; for example Shetland Line act as agents for their own ships at the Construction Jetty, as do P&O Ferries Ltd., Norscot Oil Services Ltd., and J. & M. Shearers have also provided agency services at the Port of Sullom Voe.

The Viking Piper laybarge being passed by the P&O Ferries 'St. Clair' at the South entrance to Lerwick Harbour, with the Bressay Lighthouse in the background. *(Photograph courtesy of and by Dennis Coutts)*

Rig Sedco 703 being anchored in Breiwick Bay, Lerwick. *(Photograph courtesy of and by Dennis Coutts)*

Shell tanker 'Litiopa' berthing at Sullom Voe, July 1980. *(Photograph courtesy of BP)*

Aerial view of Sullom Voe Terminal, late 1980. *(Photograph courtesy of BP)*

Sumburgh Airport with the new Wilsness Heliport in the foreground. *(Photograph courtesy of and by Dennis Coutts)*

The Drillship 'Petrel' in Lerwick Harbour in 1980. *(Photograph by Dennis Coutts, courtesy of Shell)*

Lerwick Harbour viewed from the Greenhead looking south, with Norscot Base liquid mud plants and barytes stockpiled in the foreground.
(Photograph courtesy of Norscot Oil Services Ltd.)

Chapter Seven

Telecommunications

Telecommunications have played an important part in abetting oil related developments in Shetland, and can be categorised as follows:

I Telecommunications affecting Offshore installations.

II Telecommunications affecting Onshore installations.

I Telecommunications affecting Offshore Installations

Telecommunications to offshore installations are controlled by British Telecom International (formerly the External Communications Executive, part of the Post Office), which is responsible for all telecommunications terminating in or passing through the United Kingdom to and from external sources. In 1973, the External Telecommunications Executive established a Task Force to investigate the oil industry's telecommunications needs. This study revealed that major developments would occur offshore and that conventional 'line of sight' microwave radio communications, commonly used for distances of up to 50 miles, would be ineffective for production platforms which might be situated up to 200 miles from the nearest land based link. Undersea cables were unsuitable because of a constant risk of breakage due to shallow water fishing. Satellite communications were considered, but at the time no convenient satellite existed for that form of communication to be used. Thus, a trans-horizon radio or tropospheric scatter system of communication was decided upon, whereby a high powered microwave radio signal is beamed into the troposphere, where it suffers 'scatter' from random reflections and refractions. A very small but still useable signal (one hundred million million millionth of the transmitted signal) reaches the receiving antennae.

This technology was invented by Marconi himself in the 1930's but because of the complex technical problems associated with it, particularly with regard to obtaining the necessary degree of amplification, it did not undergo any commercial application until the 1950's. The Marconi company had, by the 1970's, considerable expertise with this technology abroad, and it has now developed with the full support of the Post Office for use in the somewhat difficult climatic conditions of the North Sea.

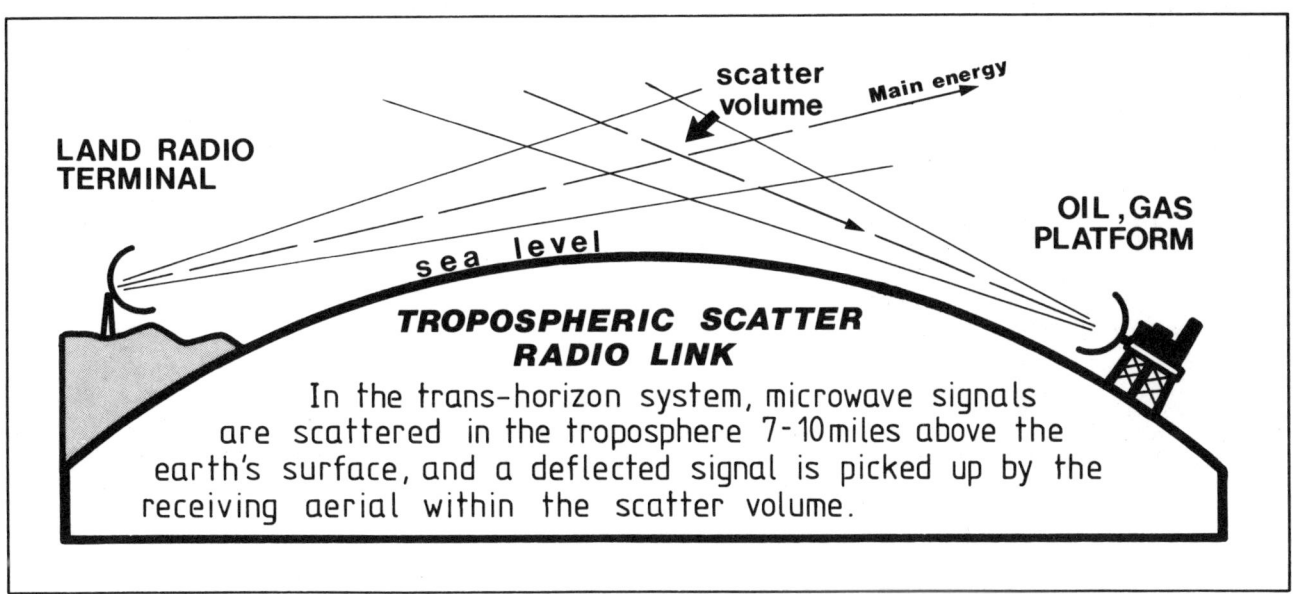

Post Office Radio Links to the Oil Platforms

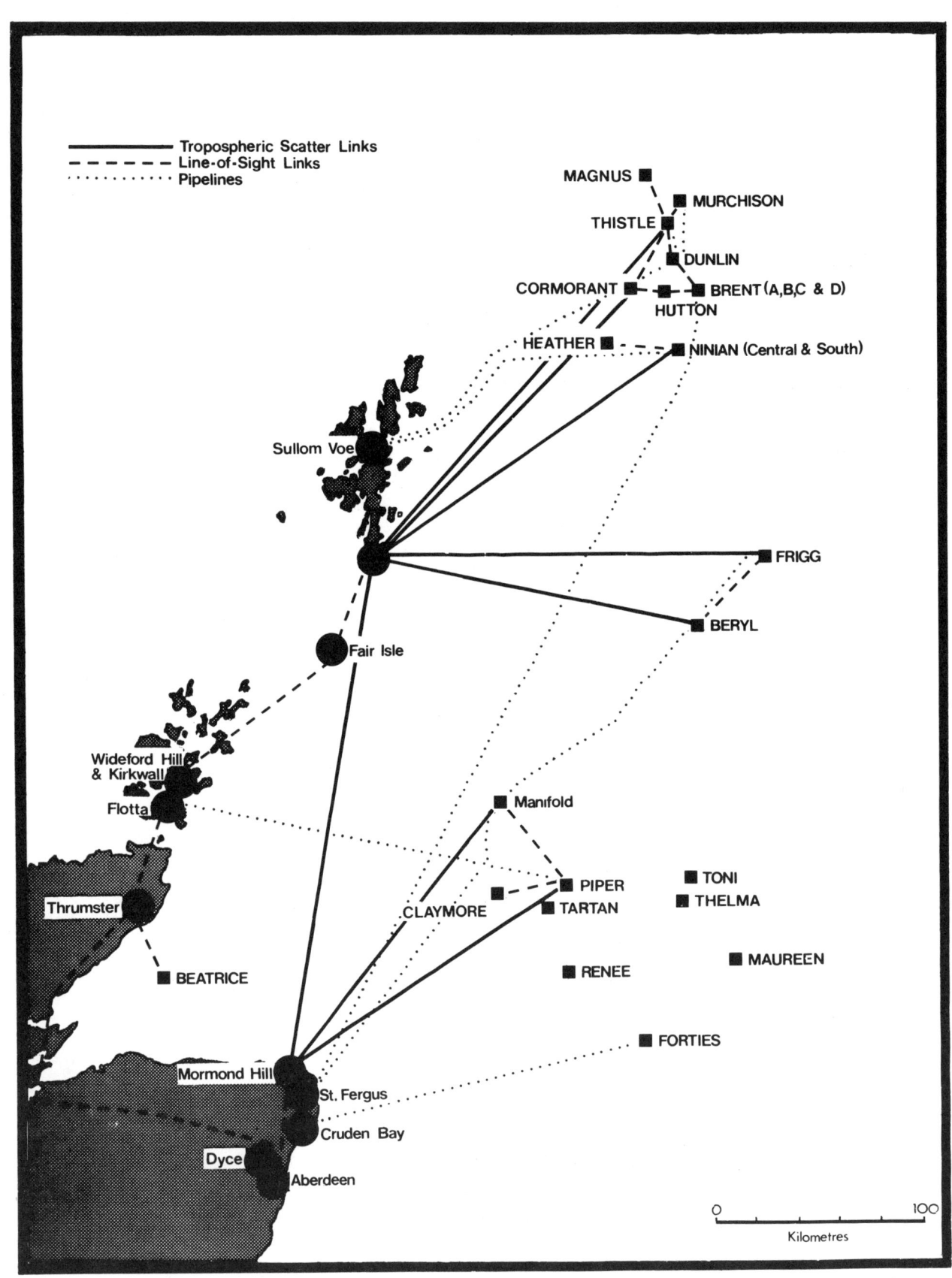

Map Courtesy of Scottish Telecommunications Board.

In early November 1973, it was announced by the Post Office Telecommunications Division, that a radio station was to be constructed near Sumburgh, Shetland, as part of a new Telecommunications network for oil and gas production platforms in the UK North Sea.

Mr Edward Fennessy, Managing Director of Post Office Telecommunications, commented at that time 'Since gas and oil exploration began, the Post Office has recognised the need for good communications between offshore and the mainland, and has been meeting the needs of mobile drilling rigs, support and supply vessels, and pipelayers since 1965. This new venture is to meet the production phase due to begin in 1975'.

The new telecommunications network was to consist of two radio stations — one in Shetland and one near Fraserburgh. By the end of 1980, a total of £7½ million had been spent at the two of these stations on buildings and equipment. These stations are equipped with giant billboard aerials to provide communications for the distant production platforms.

A service was first provided from Shetland to the Beryl platform in December 1975, Frigg Field was connected to the Shetland station by trans-horizon link in May 1977, and Thistle Field in October 1977. Temporary links were made via Thistle Field to the Brent and Dunlin Fields in the winter of 1977, using line of sight transmission. Service to the Ninian Field, and thence to Heather commenced in September 1978 (see map showing links). By the end of 1980, there were nine oilfields having links to the tropospheric scatter station in Shetland: Beryl, Brent, Cormorant, Dunlin, Frigg, Heather, Murchison, Ninian and Thistle.

Not all the platforms in the East Shetland Basin Area have direct tropospheric scatter links with Shetland. For example, when there is a concentration of oilfields, and hence production platforms, in an area, tropospheric scatter communications may be with perhaps only two platforms — A and B — which will act as 'hosts' controlling communications to other platforms in their area by line of sight microwaves. If the telecommunications are normally transmitted from the mainland to the station on platform A, under fault conditions on station A, traffic would automatically be switched to the station on platform B. If a fault arises in the middle of the system, traffic is then shared fairly between the platforms concerned. This system of operation, known as the 'alternate route switching system' was arranged by the Post Office as a method of trying to protect traffic to the platforms at all times.

The North East Shetland Basin Area Communications System (NESBACS) works on the alternate route switching principle, and is the largest UK offshore telecommunications network, involving eight platforms. All four platforms (A, B, C and D) on the Brent Field are linked together by line of sight. From Brent 'C' platform there is a line of sight link to Dunlin and Cormorant A; Dunlin and Murchison have line of sight links to a tropospheric scatter 'host' station on Thistle. Thistle also has a line of sight link to Cormorant A where the other tropospheric 'host' station is based. (See diagram).

The NESBACS system, on which Shell Expro has spent £5 million over five years, is the result of collaboration between Shell Expro, British Telecom International (formally External Telecommunications Executive of the Post Office) and Marconi Communication Systems, who designed and installed the major part of it.

The first NESBACS station to become operational was Thistle in late 1977; the other tropospheric scatter station, on Cormorant A, came into use in March 1980.

The two host stations alternate on a two weeks on/off basis, with microwave hook-ups, through Brent C Platform to the five fields and eight platforms in the system.

A great deal of critical marine navigation, aeronautic and hydrocarbon flow data passes through the system, communications, being in either digital or spoken form. As a result of this, NESBACS is equipped with a complicated back-up system, and in the event of a major generator failure at either of the host stations on either Thistle or Cormorant, traffic will be switched from one station to the other in approximately 11 seconds.

Communications are organised in groups of "channels", one channel being the equivalent of one telephone channel. Through these links an oil production platform virtually becomes a subscriber on the Aberdeen telephone exchange, having the STD and IDD facilities* that implies, being able to telex 164 countries, and dial direct by telephone 105 countries.

---

*STD = Subscriber Trunk Dialling   IDD = International Direct Dialling

Telecommunication Links
Between Shetland and the oil platforms in the north east Shetland basin

Perhaps the most interesting use of this telecommunications system is to help control and co-ordinate the platform and pipeline operations which are offshore, from onshore. This is done by each offshore installation monitoring its own operation and performance by processing information from approximately 2000 instruments. These instruments measure plant and process conditions such as: pressures, volumes, flows, temperatures and vibrations, whilst others indicate positions of valves and the state of the compressors, pumps and generators. The volume of data gathered by scanning these instruments on each platform, often as frequently as once every ten seconds, requires processing by an on-board computer.

This data is then transmitted via line of sight microwave link and tropospheric scatter to Shetland.

From Shetland the information will then be transmitted by tropospheric scatter to Mormond Hill, then by land line to Aberdeen, in the case of the NESBAC system, and also in the case of some of the other platforms in the area not in the NESBACS system, e.g. the Beryl platform. Some information is also transmitted to Sullom Voe from Shetland. The management of the pipeline systems via these telecommunications networks is a very considerable technological achievement. Several aspects of operations are monitored, including:

A. **Pipeline Integrity Systems**

    The pipeline integrity systems are designed to check that there are no leaks in the pipeline. Leaks may be caused by various events occurring, for example, a pipeline being ruptured by a ship dragging its anchor. Leaks are detected by the flow rates at the platforms and Terminal at Sullom Voe being constantly compared, along with other variables such as pressure. This information is transmitted via the telecommunications network.

B. **Platform Operations Co-ordination**

Each individual platform is monitored. There are several platforms feeding into the Brent and Ninian Pipeline Systems. The telecommunications systems are used to help co-ordinate the operations of the oilfields in order that the Pipeline Systems run smoothly.

For the Ninian Pipeline System there is a tropospheric scatter link to the Shetland station and carrier systems from there to the Sullom Voe Terminal. All data and speech is transmitted, and like the Brent System, there is computer display. There is line of sight microwave communication from Ninian Central Platform to Ninian South and Ninian Northern Platforms and to Heather, with tropospheric scatter from Ninian Central to Shetland.

Future Developments

Shell was reported in 1980 to be spending another £2 million on NESBACS to increase the number of channels dedicated to its own use.

There is also provision being made for a telecommunications service to other Fields offshore Shetland such as the Magnus Field. It is therefore anticipated that telecommunications in this form will continue to develop with the constantly changing scene offshore Shetland.

II Telecommunications Affecting Onshore Installations in Shetland

Developments onshore in respect of oil related telecommunications have been, for the major part, centred on Sullom Voe. At first, in March 1975, a mobile telephone exchange was established at Sullom Voe, there being no STD in that area at the time. There were 12 connections installed, with STD dialling to come later. In 1976/77 a small automatic exchange was installed on a temporary basis, at a cost of £170,000.

The permanent electronic exchange for Sullom Voe was commissioned on the 30th of October 1980, at a cost of nearly £1 million — £0.6 million for the linking and switching equipment plus £0.3 million for transmission equipment. A further £40,000 worth of additional equipment has since been installed. There has been a very considerable increase in the number of working connections installed at Sullom Voe, from 12 in March 1975, to 565 in December 1980.

The superseded Strowger (electromechanical) exchange has also been retained, mainly to serve the 70 pay on answer call boxes provided for the convenience of the large work force in Firth and Toft Construction Villages.

Investment in plant between exchanges must of necessity keep pace with that made in the exchanges themselves. Two new radio systems were commissioned in 1980; a 960 channel system between Sullom Voe and Scousburgh Hill and a 900 channel system from Scousburgh Hill to Shurton Hill and hence to Lerwick by coaxial cable. Together these two systems represented almost £0.75 million of investment. Substantial sums have also been spent on conventional underground plant, increasing the capacity of both local and trunk routes.

Work on a new Pulse Code Modulation (PCM) system from Sullom Voe to Lerwick — the first of its kind in Shetland — to allow simultaneous transmission of many calls over the same physical pair of wires has recently been started.

At Lerwick Exchange there have also been developments. Apart from catering for local subscribers, Lerwick Exchange is the main switching centre for trunk traffic for the entire mainland. Substantial equipment extensions have taken place in the past two years with an estimated £0.9 million having been spent in augmenting the local trunk and international dialling equipment. Further expansion is foreseen and an extension to the building is now in progress.

At Sumburgh telephone exchange £12,000 was spent on new equipment in 1980, and a further extension is in progress, with a new electronic exchange planned for 1982/83.

Developments which partially owe their pace to the impetus of the Oil Era

The development of STD in Shetland would have occurred without the oil impetus, but probably the extent of the facilities, and the growth rate of their use would have differed. In March 1975, STD was introduced into south Shetland, and in May of that year, into the Yell area. The total cost of providing this facility was £1.5 million. Of that sum £600,000 was spent at Lerwick, and £200,000 at Mid Yell, on switching centres. The introduction of STD has had several effects:

1. There has been a very considerable growth of traffic, some of which could probably be attributed to the increased access available on the STD system. From 1975 to 1978 trunk call growth in Shetland averaged about 30% per annum. This has eased in the past two years when an overall growth of 33½% was in evidence, which was still substantially greater than the mainland of Scotland where growth in the same two year period was below 20%.

   The number of trunk calls made in the calendar year 1980 by all Shetland subscribers was 3.6 million. In addition over 42,000 calls were dialled direct by Shetland subscribers to over 100 countries throughout the world.

2. There has been a change in emphasis in the staffing situation: Telecommunications staff in Shetland fall into two broad categories — Engineering and Operating. The former have gradually increased with system size and in 1980 stood at 79 in contrast to only 49 in 1973, before oil related demands on the system were felt. The latter category — the operating staff have shown a steady decline in recent years as the percentage of subscribers dialled calls has risen (currently 93.5 percent of trunk calls). From 57 telephonists — 41 full time, 16 part time in 1973 there had been a reduction to 28 — 24 full time and four part time — by the end of 1980.

   Recruitment of both engineering and operating grades in Shetland has been a problem since the Oil Era began, although the situation has eased in respect of operators. At January 1981, there were 12 engineers and three telephonists on continuous loan (on a rota basis) to Shetland from the main office in Aberdeen. British Telecom's policy is to recruit local staff but adverts and interviews have so far failed to obtain suitable staff to fill existing vacancies.

3. Lastly, there has been a considerable increase in telephones connected — from 4500 in 1975 to 7953 working lines at December 1980 — a growth of approximately 77 percent.

The influence of oil thus continues to be strongly felt, and although growth levels have now eased, British Telecoms is continuing in its efforts to improve the communications network to the benefit of both the business and residential sectors of the Shetland Community.

Chapter Eight

Oil Related Infrastructure and Social Considerations

From the developments described in the foregoing Chapters in this book, it is evident that the Shetland Islands Council, as local authority, will have had to meet considerable oil related infrastructure needs to support these developments.

Apart from the needs of the Sullom Voe Terminal Project, Lerwick Harbour and Sumburgh Airport, as commercial or industrial developments, needs have been created for the general increase in population which has resulted from these developments. In particular there has been an increase in the younger age group so that housing and education needs have grown dramatically. (See Appendix for Table XII on Population Growth).

By Infrastructure is meant:

(a) Water Supplies

(b) Sewerage

(c) Roads

(d) Housing

(e) Educational Facilities

By social considerations is meant:

(f) Social Work

(g) Community Facilities

(h) Medical Facilities

(i) Police

(j) Fire

OIL RELATED INFRASTRUCTURE

(a) **Water Supplies**

At the end of the Second World War the only public water supply schemes in Shetland were those serving the Burgh of Lerwick, and the village of Scalloway. During the quarter century after the war both these schemes were improved, and a policy of providing rural water schemes under the Rural Water Supply and Drainage Act 1944 was adopted. Schemes were provided covering most of the County, and ranged in capacity from 10,000 to 60,000 gallons per day. The water was obtained from upland catchment areas with stream or loch intakes, and treatment works comprising upward flow sand filters and simple chemical plant were installed.

The last of these schemes served the area of Mossbank, Firth, and Graven on the south side of Sullom Voe. This supply was replaced by Sullom Voe Phase I, a scheme from the same source increased by the impounding of untapped sections of catchment, and having a design capacity of some 270,000 gallons per day, compared to 20,000 gallons per day for the original scheme (see chart 'Sullom Voe Water Supply'). This was a crash scheme designed to serve the initial stages of the Terminal, and was followed by the main scheme involving taking the water from two loch catchments in the Parish of Northmavine: Eela Water; and Roer Water on the slopes of Ronas Hill. The total capacity of this scheme is some 6400 cubic metres or 1,408,000 gallons per day. Treatment of the main Sullom Voe scheme is by the sedimentation process and rapid gravity filters. These schemes were completed by the end of 1978 and cost approximately £5 million.

At Lerwick a new dam was completed in 1974, which guaranteed an adequate supply for the foreseeable future. A new water treatment works was completed in 1980 at a cost of about £1.37 million; in addition, there were new pipelines and tanks to improve the supply to the new developments in the Gremista area which cost approximately £150,000.

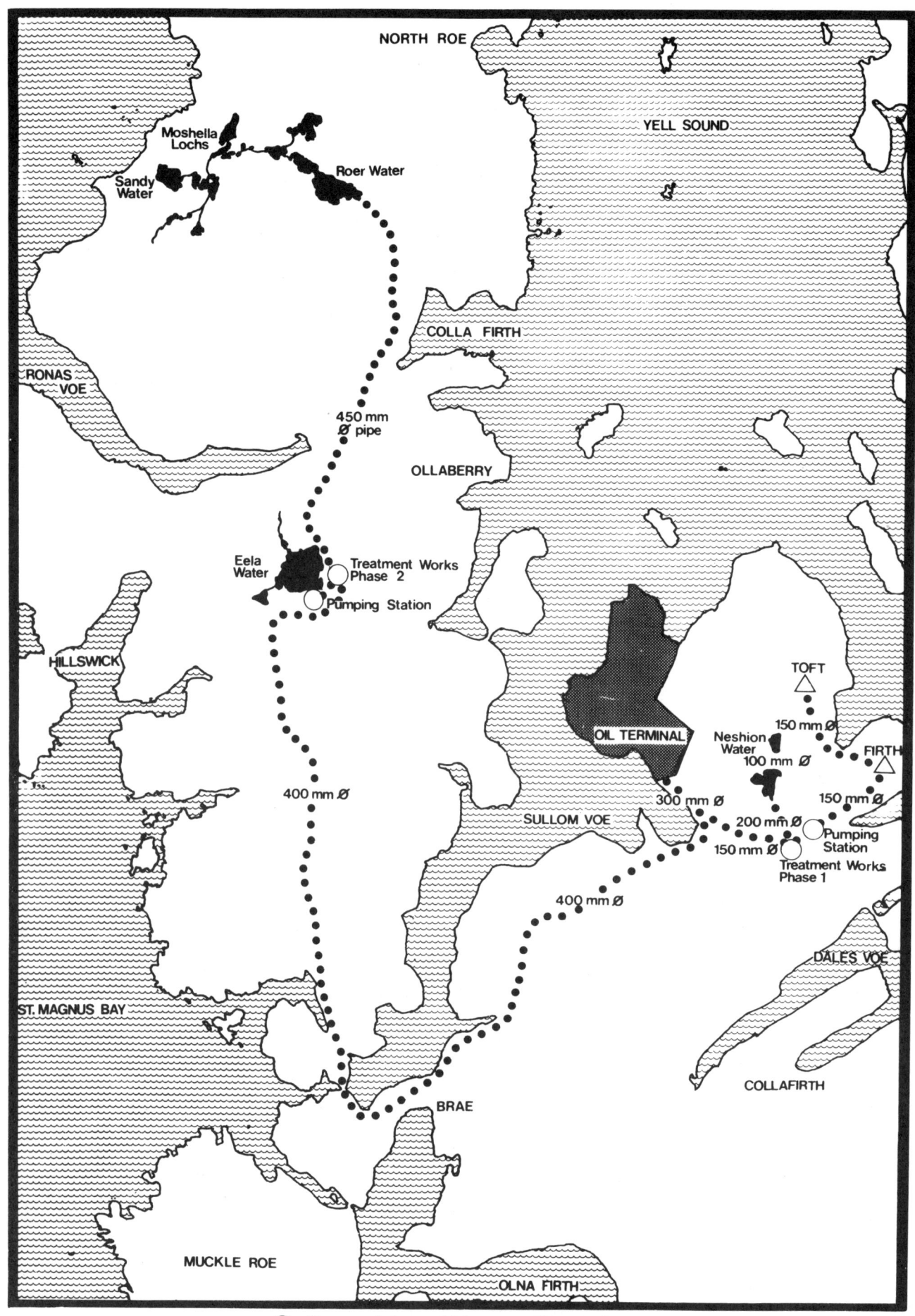

Sullom Voe Water Supply.

SULLOM VOE WATER SUPPLY

Phase 1

Startup Date: 1977.

Cost: £400,000 approximately.

Area from which water was taken for this phase: Neshion Water, Sand Water, and the Burn of Laxobigging.

Capacity: 270,000 gallons per day to be supplied to the construction villages at Firth and Toft, and the local population at Firth, Graven and Mossbank. Some of the water will also go to Calback Ness by gravity feed.

Phases 2 and 3

Startup Date: For the treatment works at Eela Water to Graven (what some have called Phase 2), 1977. For the water supply from Roer Water to Eela Water, (Phase 3), 1978. The Brae-Graven link joins with Phase 1.

Cost: £5 million approximately.

Total Water Supplied: From Eela Water, about 416,000 gallons per day. From Roer Water, about 984,000 gallons per day.

Total: 1.4 million gallons per day approximately.

Distances: Phase 1, 11.35 kilometres (from Sandwater to treatment works at Laxobigging with links to Calback Ness and Firth). *Phase 2,* 23 kilometres (Eela Water to Graven). *Phase 3,* 9 kilometres (Roer Water to Eela Water)

Contractors:

Phase 1: Gray Thomson, pipelaying
Knockbreda, civil works (e.g. building treatment plant, and tank bases).
Paterson Candy International (P.C.I.) supplied the treatment equipment.

Phase 2: Streeters of Aberdeen did the pipelaying; Knockbreda and P.C.I. did the same as in Phase 1, civil works and treatment machinery.
Babtie, Shaw & Morton, the consulting engineers, did the design work for Phase 2 (SIC did Phase 1).

Phase 3: (let with Phase 2) Streeters, and Babtie, Shaw & Morton were the companies contracted by Shetland Islands Council for this work.

provided. Currently, in 1981, a new drainage scheme is progressing in Lerwick, being built to take effluent from the new area of expansion of the town, at Sound and Clickimin, to Rova Head. A loan from the European Investment Bank has been given to help this scheme. It is envisaged that this scheme will eventually take effluent from the whole town.

(c) Roads

The main road system in Shetland was constructed during the last quarter of the 19th century and was reconstructed under the Crofter Counties Grant Scheme, between 1935 and 1941. Major improvements were begun on the road from Lerwick to Toft to improve the North Isles ro-ro ferry link before 1974, but most Shetland roads remained simple country roads. These roads floated on peat in places and had a light bitumen macadam surface, on handset bottoming, and were generally adequate for the light traffic they carried, the width varying between 3 and 5.5 metres, with turf verges and open ditching where necessary.

Early in 1974, it became apparent that the Lerwick to Sullom Voe road was suffering from the rapidly increasing traffic and the Council embarked on a policy of reconstructing the spine road from Sullom Voe in the north to Sumburgh in the south, the width adopted being 6.3 metres, with 2 metre hard verges, underlying peat being removed where economically possible. The first section of 8.8 kilometres was started later in 1974, and was substantially completed by early 1976. The second section, some 5.5 kilometres in length, was opened to traffic in early 1977. A section some 6 kilometres in length was opened in July 1978. Work is now proceeding on the Lerwick to Sumburgh section and the entire programme is expected to be completed by the late 1980's. The design and construction is in accordance with national design criteria, and is designed for a speed of 50 mph where feasible.

The Council is also engaged on a policy of improving the secondary roads linking the local communities to the main spine road, and to those roads linking the various quarries to the Terminal site. The road through the village of Brae has been totally re-constructed with a pavement along its entire length, for example. A new road was also constructed from Brae to Graven; in this instance, the oil industry shared the costs of providing the facility with the Shetland Islands Council.

To date the Shetland Islands Council has spent over £9 million on oil related road developments. (See Chart XXIV).

(d) Housing

The Shetland Islands Council strategy to meet the housing needs of incoming workers was to proceed with a dispersed type of development whereby new housing was integrated into existing settlements which already had such basic support facilities as schools, shops, churches, post offices, halls, etc.

This strategy was based on the expansion of the four communities in the Sullom Voe area, Brae, Mossbank, Firth and Voe, two of which, Mossbank and Firth, were in a state of population decline.

The first scheme of 100 houses was located within the existing village of Brae, being the community with the most immediate facilities to offer. Houses were designed in sympathy with the local style, being single one and a half storeys high, four or five person in size, and planned in small informal groupings to reflect the Shetland 'crofting' township pattern.

For speed and convenience the houses were constructed on the timber frame principle, clad externally with concrete blocks and finished in a traditional roughcast harling with black asbestos slate roofs, resulting in a simple but positive style of house.

In view of the urgency with which these houses were required, lack of contractors on the islands, and difficult site and working conditions, the completed cost of the houses was in the region of £20,000 each, this being 50 percent over the cost limit of comparable type houses on the mainland of Scotland, at that time.

The Brae scheme contained 90 percent oil related and 10 percent general needs houses. A further 24 houses were completed at Brae in 1980. Another scheme, similar in type and style, of 51 houses was constructed at Mossbank, as an 'infill' to the existing community. This scheme was 90 percent oil related, and 10 percent general needs — sheltered houses were included for local old people.

OIL RELATED ROAD IMPROVEMENTS   Chart XXIV

Date	Area	Distance* (kilometres)	Cost £
1974-76	Kames Improvements	9 km	573,900
1976-77	Brae Junction — Mavis Grind	2 km	71,300
1976-78	Camp — Brae Junction	2 km	66,680
1975-77	Grantfield — Gremista	2 km	494,120
1977-78	Gremista — Ladies Drive	1 km	267,740
1977-78	Ladies Drive — Br. of Fitch	1 km	324,240
1977-78	Br. of Fitch — Girlsta	5 km	1,207,840
1976-77	Girlsta — Sandwater	6 km	833,410
1979-81	Br. of Fitch — Scalloway	4 km	1,253,720
1978-81	Mossbank	2 km	269,630
1978-80	Sound	1 km	272,890
1975-77	Levenwick	2 km	248,770
1978-79	Gulberwick	5 km	1,994,240
1978-81	Clumlie	1 km	382,420
1978-81	Northmavine	5 km	899,660
1977-80	Gremista — Norscot	2 km	38,270
	Total	50 km	£9,198,830

*Distance is rounded to the nearest kilometre.

Note: In addition another 29 kilometres of road improvements have been undertaken which are not strictly oil related namely:

Early 70's	*Tingwall/Bixter/Aith*	*10 km*
1974-81	*Yell*	*17 km*
1972-73	*Unst*	*2 km*

(b) Sewerage

The Shetland scene regarding sewerage consisted of isolated schemes with either simple settlement or direct outfall to the sea. The remaining areas had no public drainage and houses were served by private septic tanks. Complete new sewerage schemes for the new housing, schools and other developments have therefore had to be

SULLOM VOE AREA

Chart XXV

OIL RELATED HOUSING DEVELOPMENTS (As originally mooted in the 1974 Sullom Voe District Plan)

Village	Population 1971	Estimated Annual Increase per year 1975-80	Estimated Houses built by 1980	Estimated Population 1980	Estimated eventual No. Houses to be built	Estimated eventual Population
Voe	220	10 families per year —30 people	50	370	125	600
Brae	320	35 families per year average	175	820-920	275	1120-1220
Mossbank	120	25 families per year average	125	420-520	250	800-900
Firth	6	50 families per year average	200	550-600	350	1000+

HOUSING IN THE SULLOM VOE AREA (As at February 1981)

Scheme	Location	No. of Houses	Date Occupied
1	Brae	100	1976
2	Mossbank	51	1977
3	Voe	15	1977
4	Firth	100	
	Phase 1	43	1977
	Phase 2	57	1978
5	Brae	24	1980

A scheme of 100 houses followed on at Firth, grouped along the side of a voe, where a once thriving but now completely derelict crofting community had existed. These houses are 100 percent oil related, and have been set in tight knit groups terraced into the steep hillside. A small scheme of 18 general needs houses has also been built at the village of Voe, but several of the houses contain people in oil related employment.

Housing for the Delting area was based on the needs identified in the Sullom Voe District Plan, which was prepared for a period of five years covering 1975/80. The plan acknowledged that the rate of development in the area would be dependent on the progress of the construction of the Terminal. It was suggested that 540 houses would be required by 1980. Up to the present time, 455 houses have been completed, leaving a shortfall of 85. This figure, however, excludes 50 temporary houses constructed at Firth by BP.* When these are added to the houses already built, this gives a total of 505, 35 short of the original plan. The current Shetland Islands Council financial plan allows for a further 38 houses up to 1985 and 91 in the post 1985 period, whereas the Sullom Voe District Plan suggested a further 450 houses for this period. The reason for this difference is the current lack of any specific downstream petrochemical proposals.

*Many of the private houses were built by BP. See Chart XXV and Chart XXVI "Sullom Voe Area Oil Related Housing Developments" and Sullom Voe District Plan — Housing Plan.

SULLOM VOE DISTRICT PLAN — HOUSING PROVISION Chart XXVI
1975 to 1980

Properties Completed by 1981

Brae	Local Authority —	134
	Private —	54
Voe	Local Authority —	18
	Private —	37
Mossbank	Local Authority —	51
	Private —	27
Firth	Local Authority —	100
	Private —	34

Total Completed Properties   Local Authority — 303
                                      Private — 152
                                              ___
                                              455

SCHEMES PLANNED

Scheme	Location	No. of Houses
6	Voe (Phase 2)	20 — design stage; 1981/84
7	Brae	26 — 1984/86 25 — 1986+
8	Mossbank	35 — 1985+ 25 — 1986+

The house building figures have therefore kept reasonably on target and this is borne out by the population figures. The total population in Delting, has increased by 1500 since the Oil Era began, which corresponds closely to the total predicted increase of 550 families, i.e. 1650 people.

Although the Sullom Voe area has been probably the most oil affected in Shetland, other areas have also been affected and had housing needs. Perhaps the most interesting feature of a comparison in the provision of Council housing from 1971 to 1981, is the tendency to move away from Lerwick, and for Council housing developments to be spread through the islands. For example, in 1971, there were 987 Town Council houses in Lerwick, and 391 County Council houses in the rest of Shetland — the total Council housing stock being 1378. In contrast, by 1981, not only had the Council housing stock nearly doubled — reaching a total of 2475 houses — but there were 1286 houses in Lerwick and 1189 in the country, suggested that the balance between town and country Council housing had been redressed completely. Private housebuilding had also greatly increased, and almost equalled the amount of Council houses built over the period.*

Over 60 percent of the houses in Shetland are privately owned.

(e) Educational Facilities

As a direct result of the increase in population due to incoming oil related workers and their families, the existing old schools and educational facilities throughout Shetland (not just the Sullom Voe area) became totally inadequate.

A school building programme to meet the needs of all grades of education was embarked upon. Schools have since been constructed in the oil affected areas — Brae, Mossbank, Lerwick, Cunningsburgh, Dunrossness and Whiteness — at a cost of over £9 million (see Chart XXVII "School Building Programme 1975-1981").

These developments reflected a growth in primary school pupils from 2014 in 1971 to 2738 in 1980 — an increase of almost a third in the number of children to be catered for; and at the secondary school level an increase from 897 in 1971 to 1619 in 1980 — a growth rate of almost 100 percent. The number of teachers required grew accordingly, from 179 in 1971 to 308 in 1980.

SCHOOL BUILDING PROGRAMME 1975-1981 — Chart XXVII

Schools Completed (to February 1981)

	Cost	Size	Date completed
Brae Primary School	£950,000	7 classrooms	1976
Cunningsburgh Primary	£250,000	4 classrooms	1978
Dunrossness Primary Extension and Swimming Pool	£250,000	4 classrooms	1978
Whiteness New Primary	£250,000	4 classrooms	1978
Sound New Primary and Community Centre, Lerwick	£650,000	12 classrooms	1978
Sound Nursery Wing	£9,000	2 rooms + ancillary	1979
Mossbank New Primary	£1 million	10 classrooms	1979
Anderson High School and Hostel Accommodation	£6 million	70 classrooms	1980
Hamnavoe School	approx. £400,000	5 classrooms	1980

Under Construction

	Cost	Size	Date completed
Brae Secondary School	£1.4 million	for 360 pupils	To be completed 1981
Happyhansel Primary School	£500,000	3 classrooms	January 1982

Plans are completed for schools at Aith and Sandwick and a new school is being planned at Scalloway.

Summary: 9 projects completed at a cost of £9,759,000 approximately.
2 projects under construction at a cost of £1.9 million.
3 large projects yet to come.

SOCIAL CONSIDERATIONS

(f) Social Work

Shetland Islands Council has a statutory obligation under the Social Work (Scotland) Act 1968, to provide social services for the community.

Many of the services provided by the Social Work Department have been under the terms of the 1968 Act. For example, since 1972, 23 schemes of sheltered housing for old people have been completed in Shetland — the smallest being of five units, the largest 12. Sheltered housing policy is to provide a small number of units in order to obtain a good age mix in housing schemes and thus avoid ghettos of one particular age group. A 30 bed old people's home at Kantersted in Lerwick replaced a smaller unsuitable house at Leog, which is now a Home offering places for up to 12 children.

A new day centre for the mentally and physically handicapped has been built at Lerwick, offering 50 places on a day care basis, and includes a stroke and fracture rehabilitation unit in consultation with the nearby New Gilbert Bain Hospital. For the elderly there is a meals on wheels service throughout Shetland, and a domiciliary laundry service.

There is also a hostel for the mentally handicapped in Lerwick, currently offering six places.

The above noted developments would probably have occurred without the Oil Era, although there might have been fewer housing schemes in which to include sheltered housing. There have been however, considerable demands made on the social services resulting from the social changes wrought by oil.

The Social Work Department has increased its staff, and put one full time social worker at Brae to serve Delting and Northmavine areas, and another in the Northern Isles (Yell, Unst and Fetlar) which have also experienced changes.

The rate of "referrals" — people with problems who are referred to the Social Work Department — has been rising steadily since 1972. The referral rate from the incoming oil related population has been as high as 10 percent, for example, in the Dunrossness (Sumburgh airport) area. Within this there has been the entire range of social work — arranging adoptions, home helps, dealing with delinquency, cases under the Mental Health Act, arranging for incomers aged relatives to be moved to nearby Homes in Shetland, etc.

Whilst the increased affluence in Shetland has brought many positive features, the social workers have had an increasing number of cases of alcohol abuse and children neglected by working mothers to deal with.

The hospital has also been busier with oil related work which has meant greater pressure to get the older patient out more quickly. A shortage of nurses (partly due to lack of reasonably priced accommodation) has meant the geriatric hospital is not fully manned, and hence fewer beds are available there. There has also been greater pressure on the home help service as people pursue more lucrative oil related jobs. People also seem to have marginally less time for the elderly — for example, to ensure they get their peats cut and cured.

There has been oil related money given to social work projects from the Charitable Trust, although the main source of finance is acquired through statutory means.

The former British Legion building in Lerwick was purchased with money from the Charitable Trust to be used as a day centre and luncheon club for the elderly. It is also widely used by the local community for functions in the evenings and running costs are now funded by statutory sources.

Charitable Trust monies have been dispensed by the Social Work Department both for community projects, and for allocations to needy individuals or families.

Community projects have included:

(i) giving money to the Shetland Health Board for an experimental 'night sitter' (night nurse) service, to allow patients to remain at home;

(ii) supporting Red Cross House in Aberdeen as a place to stay for Shetland patients, and their relatives visiting them in hospital;

(iii) providing wheelchairs at Sumburgh Airport;

(iv) giving substantial support to the playgroup movement;

(v) acquiring two specially adapted coaches for the disabled;

(vi) giving cash to local hospitals to improve amenities, e.g. colour television;

(vii) converting a boat to allow the disabled to fish.

Projects to support individuals and families in need have included;

(i) £8500 to a family with a severely handicapped child to extend their house;

(ii) specially adapted Volvo cars to allow the disabled to be mobile;

(iii) workshops for a young blind man;

(iv) holidays on the Mainland for the disadvantaged;

(v) renovations and improvements to the property of the elderly and housebound. This has been particularly useful where people had insufficient means to take advantage of statutory grants and install basic amenities.

(g) Community Facilities

The Shetland Islands Council employs community work staff within its Department of Leisure and Recreation, which was established at the re-organisation of local government in 1975. This is unique amongst local authorities in Scotland in having its community work functions within a Leisure and Recreation Department, and not under the auspices of Education.

When the Leisure and Recreation Department was first established in 1975 the scope and value of community work in Shetland had still to be realised. However, the massive demands being made on all local authority services at that time led to the emergence of a vital community work function from 1975 to 1978. It was money from the Oil Distrubance Fund that allowed Shetland Islands Council to respond to local initiatives whereby the foundations of many community work projects were laid throughout Shetland.

Since 1979 many development grants have been made to 'parent' organisations. For example, rather than the Shetland Islands Council purchase and lend out chess equipment such as chess clocks, the Council suggested to the people involved at the time to form a parent body — The Shetland Chess Association — which is able to receive funds, and the Association is encouraged to use these to secure development progress in many ways, e.g. by providing and overseeing the use of equipment for club use.

In 1981, now that statutory resources are being used to support community groups and initiatives, the Oil Disturbance monies are not used by the Department in the same way as previously. For example, the Charitable Trust has been used to give a grant to a capital scheme in the Out Skerries. To build a new hall under statutory conditions, the people of Skerries (population 96) would have required to raise £15,000 through voluntary fund-raising efforts. This would have taken some years, especially as it is not easy for a remote island to attract in people from outside their own community for money-raising. Rising building costs would have added to the difficulties. A grant of £7,000 was made from the Charitable Trust, and the Skerries people raised £8,000 to reach their target of £15,000, which it is now hoped will be matched by money from various statutory sources. Several other hall schemes have been augmented by Charitable Trust Loans on most favourable terms since 1974.

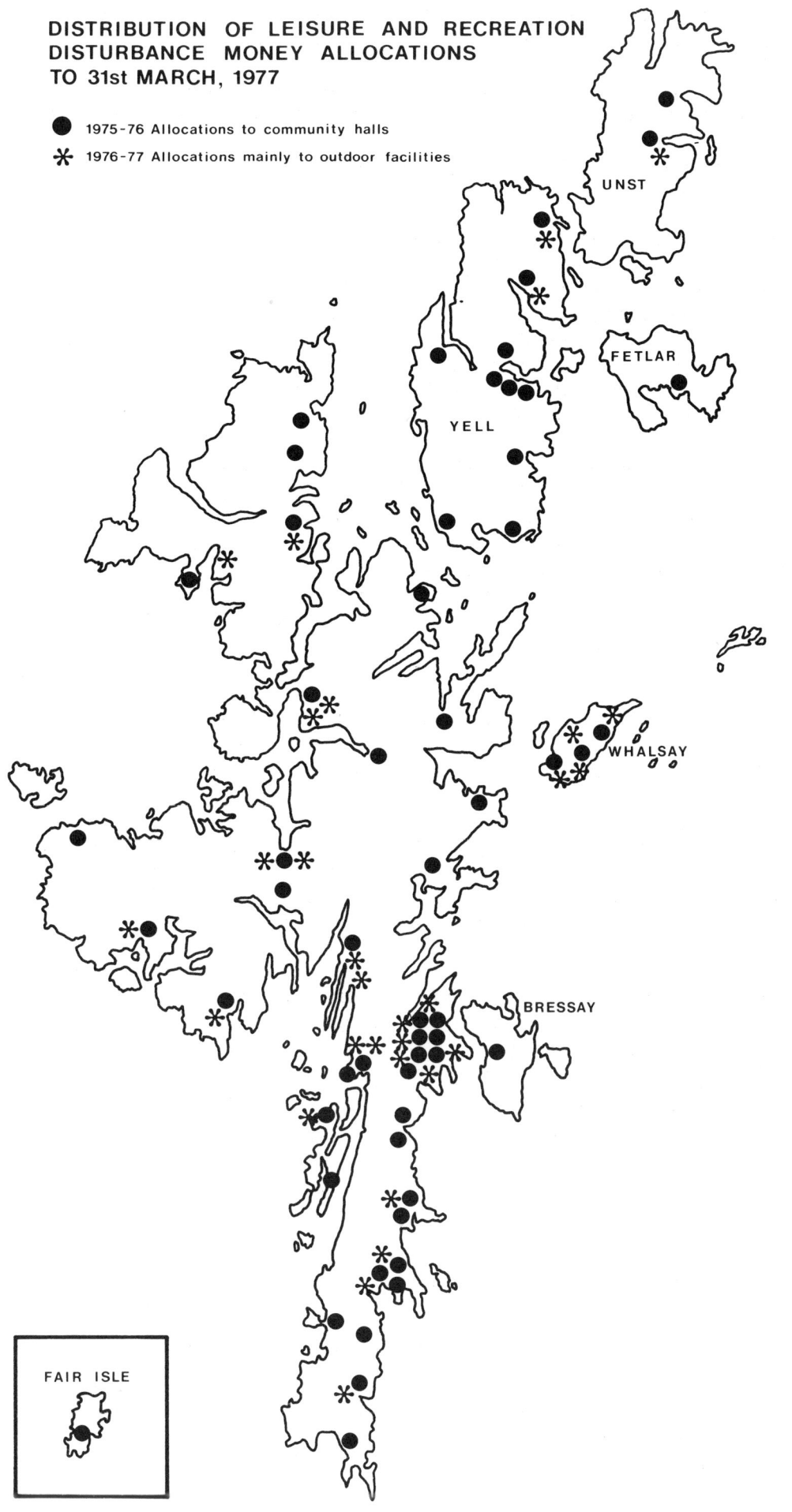

For example, Tingwall Hall Committee received a £20,000 loan at ½ percent interest with a capital repayment delayed for three years, and thereafter repayments to be over five years.

It can thus be seen that the Oil Era, while undoubtedly contributing to some difficult situations, has had positive side effects, in boosting the finance available to support expanded human resources within the Community.

In general, the policy in the use of the oil monies was to achieve as much as possible for the community by adequately supporting voluntary initiatives.

In 1975/76, £50,000 was used entirely to put into motion a scheme to improve buildings for community use, such as village halls. To quote from a report, sponsored by the Highland and Islands Development Board* "The best example of a broadly-based and successful approach to the enhancement of village halls is the Shetland scheme. Admittedly it has been financed by oil revenues, which provided grants of up to £1000 or 75 percent of the cost (whichever was the less) for agreed schemes of improvements. Nevertheless, while the Shetland Islands Council is clearly in a special position to take an initiative of this kind, the success of the scheme has relevance for grant-giving schemes in general and highlights the necessity of considering physical improvements and community well-being as part and parcel of the same problem.

"It was decided to offer grants to organisations responsible for buildings available to, and used by, young people for sporting, social or recreational purposes. The scheme attracted a widespread response, with those responsible for the majority of village and community halls in the Islands seeking grants, through a process which merely required a short questionnaire to be completed and submitted directly to the Leisure and Recreation Department. In all, 54 schemes were approved involving a wide range of improvements, and although the nature of the grant precluded any major capital works, a striking range of structural changes, improvements to fabric, acquisition of equipment of various kinds and infrastructural improvements was proposed; these have nearly all been implemented.**

"By-products of the scheme have been the resuscitation of numerous community associations that have been defunct for many years, the revival of local leadership and the identification of potential new leaders. Apart from these benefits to the community development, another feature of the scheme has been the obvious value for money obtained by the use of local labour and craftsmanship on projects that might not otherwise have been economically viable."

In 1976/77, a further £50,000 was used from the Disturbance monies to achieve improvements on outdoor facilities, again using voluntary organisations at every opportunity. During this time practically every settlement in Shetland was visited (by Leisure and Recreation Department staff). This work was continued in 1977/78, with a further £60,000 from the Disturbance monies, by which time its value was established beyond doubt.

From 1979/80, it was decided to use statutory resources**and £110,000 was incorporated into the Revenue account of the Leisure and Recreation Department for the furtherance of community work, this being increased to £250,000 for the year 1980/81. An interesting comparison in investment in community work is that in November 1980, the Minister of Health and Social Work announced two new schemes to develop local voluntary action for the whole of Scotland. This involved three grants — one of £25,000 per annum for three years, the other two of £20,000 per annum for two to three years.

---

*See TRRU Research Report No. 37 'Recreation in the Highlands and Islands — Phase 2 Regional Perspective — A commentary" published April 1978.
**See Table XI "Shetland Islands Council — Aid to Youth Facilities Project".
***See Local Government (Scotland) Act, 1973; Sections 90 and 91.

Table XI

SHETLAND ISLANDS COUNCIL — AID TO YOUTH FACILITIES PROJECT

The Village Hall

Type of Improvement	% of Successful Applications
General main hall structural improvements, e.g. timber cladding	46
Heating — installation or improvement	41
Kitchen equipment	41
Lighting — installation or improvement	28
Furniture	28
Entrances and windows — improvement	26
Storage space	22
Fire protection equipment	20
Toilets — installation and improvement	19
Insulation and ventilation	19
Decoration	19
Interior repairs	19
Kitchen — installation and improvement	17
Stage — installation and improvement	11
Roof repairs	11
Music equipment	11
Miscellaneous	9
Extensions and cloakrooms	7
Electrical rewiring	7
Sports equipment	7
Shower(s) — installation and improvement	4
Parking facilities	4

*Total applications (successful): 54

(h) Medical Facilities

The provision of medical facilities in Shetland is the responsibility of the National Health Service.* The Shetland Health Board consists of a group of people appointed by the Secretary of State for Scotland for their personal qualities or contribution that he feels that they might make. They are a representative group of people from the voluntary organisations, the Trade Unions and the Shetland Islands Council, and anyone else that the Secretary of State feels warrants the job. The people who are there are appointed on a personal basis and not to represent any political party. Backing up the Health Board, there is a permanent Secretariat, and administration.

There are four hospitals in Shetland, all in Lerwick. Three of these hospitals — the Brevik (built in 1878), the old Gilbert Bain (built in 1901), and the Montfield (built in 1928), are for geriatric patients, having a total of 87 beds; the fourth, the New Gilbert Bain, which was opened in 1961 by the Queen Mother, is mainly for general surgery, general medicine and obstetrics, having a total of 73 beds.

From 1971 to 1981, there has been an increase of over 5,000 people registered with local General Medical practitioners — from approximately 17,000 to approximately 23,000 people. Within the same period of time, there was an increase of over 200 percent in the number of casualties attending the New Gilbert Bain Hospital — from 1082 in 1972 to a peak of 3134 in 1978.** The hospital itself has not increased in size, and has suffered from staff shortages. For example, in 1980, the nursing establishment was on average 80 percent of what was needed. The increase in casualties can probably be attributed mainly to the general increase in population in Shetland onshore, and also to some offshore oil related workers coming to Shetland, as the nearest medical facility.

There has been a rise in the babies born at the hospital, from 265 in 1975 to 312 in 1979. The hospital has visiting specialists for dermatology, ear, nose and throat, opthamology, and some other fields of medicine, but for many complaints, it is necessary to go to Aberdeen, as there are only specialist facilities for general surgery (but not for other treatments) in Shetland. This is also one of the reasons why there is a shortage of nursing staff, as the Hospital is not a recognised training school, as it is too small to offer all the facilities for training, and Shetlanders wishing to train as nurses have to go to the teaching hospitals in Aberdeen and elsewhere in Scotland. Another reason has been the high cost of accommodation, as nurses cannot afford to compete with the oil inflated rents.

There are plans to improve the New Gilbert Bain Hospital from the years 1983-1993, in a three phased plan, each phase lasting approximately three years. Improvements will consist of an additional surgical ward, a new laboratory, new physiotherapy facilities, etc., as many of the areas in the hospital are totally inadequate to cope with the present needs. The site is very limited, with the result that each phase has to be completed before the next can commence, hence the time factor involved.

There are also plans to develop the Montfield Hospital, and to take some of the pressure off the nursing service by removing all geriatric beds to the Montfield, and provide a day hospital for the aged. The building programme is planned from 1981-1983.

There have been improved facilities in Brae, due to the increase of population there (mainly oil generated). In early 1981, a new Health Centre was opened there. Two General Medical Practitioners are based there, and there is a dental suite for the new dentist, who is due to start work there in 1981. There are also facilities for Health education, chiropody, physiotherapy, medical social work and nursing services. It is hoped to encourage visiting consultants from Aberdeen to visit Brae, and to develop Family Planning services, parenthood and maternity relaxation classes there.

The Health Board has also, in conjunction with the oil industry, and all the emergency services, been involved with developing a major accident and emergency plan for Shetland. This plan was revised in light of the tragic accident at Sumburgh in July 1979, which cost 17 people their lives, when an aircraft on charter to the oil industry crashed on take-off.

The Health Board relies on the close collaboration with the Red Cross, and many other voluntary organisations to help provide comforts for the patients, and provide such help as a coffee bar. There has also been established recently a League of Friends for the Shetland hospitals, and there is much active community involvement within the framework of these organisations, which have been good at raising money to help the hospital service. It has been a recurrent problem for a small Board like the Shetland one to provide finance at the right time to replace major items of equipment. To carry out capital projects of any size, the Board is dependent on the allocation of sufficient resources and additional finance from the Scottish Home and Health Department.

Through the auspices of the Shetland Health Board.
**In 1980 casualty figures had declined slightly to 2953.*

Nearly three quarters of hospital and community current expenditure goes on salaries and wages, and the proportion is rising slightly over time. It is clear that it will be difficult to expand the total health service staff at a rate much quicker than that set for total health service expenditure, thus the problem of understaffing looks likely to remain for some time. It is perhaps of interest to compare how financial resources have been distributed between the different groups in Shetland, compared to Scotland as a whole:

	Shetland %	Scotland %
Maternity	4.2	5.5
Child Health	5.5	5.4
Elderly	28.4	9.5
Mental Illness and Handicap	0.3*	14.7
General and Acute Hospitals	30.6	35.3
Primary Health Care	28.7	23.6
Other	2.3	6.0
	100.00	100.00

(i) **Police**

The Police Force in Shetland is within the Northern Constabulary which has its Headquarters in Inverness. Police services are paid for by the local authorities from the rates.

The Police Force in Shetland has doubled in size from 21 in 1971, to 42 in 1979. From 1972 to 1979 crimes and offences reported to the Police have risen over 250 percent from 578 to 2058. The main category under which crimes have risen is "Class 7" — miscellaneous crimes — such as road traffic offences, litter, etc. — where crimes and offences have risen from 432 in 1972 to 1751 in 1979. In all other categories there has been a rise in crime, there perhaps being a parallel with the situation which arose in Easter Ross when it became oil affected. The encouraging factor in this situation in Shetland, is that there is a very high rate of detection. For example, of the 77 crimes and offences against the person in 1979, 77 (i.e. 100 percent) were detected.

Presently, of perhaps greatest concern to the Police, has been the rise in drink related offences, and of accidents. (See statistics).

The Police establishment has met these new challenges, and new Section Stations have been established at Brae and Dunrossness to cope with the increased population due to the Oil Era, in addition to the Lerwick force being increased.

Most patients with these problems are treated in Scotland, as there are no facilities for them in Shetland. It can thus be seen that the pressures that the Oil Era has placed on the Health Service in Shetland are not inconsiderable, but that they have been compounded by the problems created by a fairly high percentage of an aged population needing medical help.

Accidents and Drink Related Offences known to the Police in Shetland in 1980

1. Accidents

Total No. of Accidents	320
No. of Fatal Accidents	1
No. of Persons Seriously Injured	82
No. of Persons Slightly Injured	96
Total Casualties	179

2. Drink Related Offences

No. of Drink Related Driving Offences	183
No. of Cases Involving Drunkeness	102

(j) **Fire**

Fire services in Shetland come under the auspices of the Northern Fire Board, which is based in Inverness. Fire services are paid for by the local authorities from the rates. The Northern Fire Board therefore consists of two members from each local authority in whose area it functions, with a Chairman elected from within the group. There is an executive branch of specialist fire officers who carry out the policy decisions of the Northern Board, the Chief Executive being the Firemaster in Inverness.

Since the Oil Era began, there has been a small increase in the establishment of the Fire Unit at Lerwick, which has 20 retained Firemen* plus two full time officers, and at Brae, which has now 18 retained Firemen.** The oil related airports — Sumburgh, Scatsta, Tingwall and Unst — have their own fire services, as does the Terminal at Sullom Voe.

In Shetland the fire services have to be self contained, being in an island situation, and unable to call on help from neighbouring brigades in case of major fires. This fact has affected the manning levels of the fire services, and possibly other considerations such as equipment. There are presently six fire appliances in Shetland for the fire services — three at Lerwick, two at Brae, and one at Sumburgh.

There has been a considerable increase in the number of fires attended, from 41 in 1971, to 109 in 1979, and one could attribute this increase to the increase in population.

There have also been some changes in procedures — for example, since 1979 there has been no siren going off every time there is a turnout in Lerwick, as a localised pager system (effective for a two mile radius or more) has been introduced.

From the expansion noted here, it can be seen that like practically every other service in the community, the fire services had not remained unaffected by the Oil Era!

---

*A retained Fireman is a Fireman paid an annual retainer fee for his services.
**Until 1979, there were no retained Firemen at Brae — only volunteers.

Chapter Nine

The Economic Implications for Shetland of the Oil Era and its Associated Developments

In 1971, before the Oil Era began in Shetland there was 3.8 per cent unemployment. Against this backdrop one had a reasonably buoyant economy based on the traditional industries of fish catching and processing, agriculture and knitwear.

A study made of the situation five years later, in 1976, when there was 3.5 per cent unemployment, revealed that radical changes had occurred in the structure of the economy. There had been losses of labour from the traditional industries to new (and in some cases temporary) oil related employment, which offered higher wages.

A general idea of the effect of the Oil Era in 1976/77 was that Shetland's per capita income was estimated at £1631, compared to an estimate of £1054 in Scotland, and £2009 in Britain as a whole.*

In 1971, Shetland's per capita income was 69 per cent of that in Britain — this figure had risen to 81 per cent by 1976. It was further estimated that the total income of Shetland households in 1976/77 was £31.2 million. Converted to 1971 monetary values, this would be £15.3 million. The actual total income for Shetland in 1971 was £10.1 million, representing a growth rate of approximately 8.5 per cent per annum during the intervening years.

Looking at the overall economic situation, it was estimated that Shetland had a real economic growth rate of 8 per cent per annum from 1971 to 1976/77 in terms of Gross Regional Product,** compared to a rate of 1.25 per cent in Britain as a whole, and of 1.5 per cent in Scotland, on average, over the same period. By 1976/77 oil construction and oil supply bases ranked first and fifth in terms of gross output, although the largest industrial exporters in absolute terms were ranked as fish processors (£6.9 million), oil supply bases (£6.6 million) and textiles (£2.1 million). No other industry exported over £1 million worth of goods.

Meanwhile, from 1971 to 1976 there had been an explosive rise in the cost of living, not only from the fact that there was beginning to be felt the effects of the inflation which was a phenomenon of the Western World, but also because of the advent of the oil industry and oil developments in the community.

---

*See I. H. McNicoll and G. Walker's work 'The Shetland Economy 1976/77: Structure and Performance' completed in July 1978, and subsequently published by the Shetland Islands Council.

**Gross Regional Product is the wealth produced by the economy from the start to the finish of the year.

SHETLAND INPUT – OUTPUT TABLE, 1976/77

£'000 at producers' prices

PURCHASES FROM \ SALES TO	AGRICULTURE	FISHING	QUARRYING	FISH PROCESSING	TEXTILES	SHIP REPAIR	OTHER MANUFACTURING	CONSTRUCTION	UTILITIES	TRANSPORT	DISTRIBUTION	PROFESSIONAL SERVICES	OTHER SERVICES	COMMUNICATIONS	LOCAL GOVERNMENT	OIL SUPPLY BASES	OIL CONSTRUCTION	HOUSEHOLDS	INTERMEDIATE DEMAND *1	CENTRAL GOVERNMENT	TOURISM	INVESTMENT	EXPORTS	UNREQUITED RECEIPTS *2	GROSS OUTPUT
AGRICULTURE	68.7	10.6	–	–	–	–	33.3	–	–	–	–	22.3	–	–	2.3	–	–	589.2	795.1	–	–	587.5	846.4	–	2229.0
FISHING	–	–	–	3344.5	–	–	3.5	–	–	–	–	4.2	10.5	–	–	–	–	55.8	3418.5	–	–	–	289.4	–	3707.9
QUARRYING	–	–	–	–	–	–	54.9	159.3	–	–	–	–	–	–	–	1.4	572.8	–	788.4	4.8	–	89.6	13.4	–	896.2
FISH PROCESSING	–	–	–	115.3	–	–	36.4	–	–	–	–	–	–	–	–	–	–	62.5	214.2	–	–	103.0	6946.0	–	7263.2
TEXTILES	–	–	–	–	1.4	–	3.7	–	–	–	–	–	–	–	–	290.9	–	38.9	44.0	10.4	91.6	82.8	2125.5	–	2354.3
SHIP REPAIR	–	461.1	–	–	–	–	–	–	–	146.6	–	–	–	–	–	290.9	–	–	898.6	–	–	28.9	–	–	927.5
OTHER MANUFACTURING	–	29.5	0.6	7.5	6.4	0.9	192.8	39.6	7.0	6.2	34.8	43.7	25.2	0.5	20.3	5.3	179.9	695.0	1295.2	9.9	119.0	17.8	337.8	–	1779.7
CONSTRUCTION	204.6	79.6	105.8	10.0	4.7	122.3	20.7	122.2	15.3	106.7	66.0	63.9	10.4	1076.6	143.5	114.7	237.3	2394.3	320.5	–	2994.9	–	–	5709.7
UTILITIES	27.2	47.7	98.2	8.3	2.7	13.4	14.0	8.5	38.0	82.8	149.2	76.0	5.6	219.2	35.3	107.5	607.0	1540.6	37.1	–	84.4	–	–	1662.1
TRANSPORT	6.5	101.7	117.7	79.8	19.3	7.2	24.2	541.0	10.9	460.6	1275.0	82.0	66.2	17.1	463.8	411.5	944.7	360.3	4989.5	43.1	2.3	25.6	929.5	–	5990.0
DISTRIBUTION	178.3	45.0	14.3	195.3	10.9	9.1	50.3	201.7	38.9	53.2	3.0	44.2	189.3	0.8	13.3	2.4	1356.5	4094.9	6501.4	49.7	74.4	70.5	–	–	6696.0
PROFESSIONAL SERVICES	70.0	13.8	–	27.1	13.1	2.7	36.6	25.0	12.8	339.2	64.2	192.9	93.7	45.0	2908.9	61.1	3.7	120.4	4030.2	1773.5	–	–	–	–	5803.7
OTHER SERVICES	28.5	–	12.8	–	8.1	–	15.3	47.5	2.6	14.5	42.9	40.0	208.6	3.4	83.7	218.7	726.4	3313.8	4766.8	348.3	1006.1	354.5	–	–	6475.7
COMMUNICATIONS	3.5	–	21.7	–	47.8	5.0	8.3	7.7	6.1	37.7	85.6	56.5	51.8	7.4	9.4	51.2	149.0	136.1	685.5	35.7	0.8	–	29.2	–	751.2
LOCAL GOVERNMENT	4.7	27.3	4.6	13.9	5.3	4.2	5.7	24.7	29.4	34.3	42.2	68.2	62.0	5.1	37.9	93.5	75.4	601.2	1139.6	7503.2	–	–	–	–	8642.8
OIL SUPPLY BASES	–	15.8	–	–	–	–	–	–	–	–	–	–	–	–	–	–	–	–	15.8	–	–	–	6562.0	–	6577.8
OIL CONSTRUCTION	–	–	–	–	–	–	–	–	–	–	–	–	–	–	–	–	–	–	–	–	–	25817.4	–	–	25817.4
HOUSEHOLDS	1328.8	2133.5	219.7	1088.3	1024.0	233.8	480.7	2814.0	314.9	1979.2	3475.3	3407.0	1180.0	537.5	1359.6	1540.5	1115.6	75.3	24307.5	5130.0	89.2	–	–	–	31247.4
TOTAL INTERNAL	1920.8	2838.3	484.9	5110.2	1154.6	270.3	979.8	3996.8	443.3	3124.8	5212.5	4176.2	2095.9	632.8	6195.0	2855.1	5346.2	10987.7							
IMPORTS	672.6	497.0	190.7	2042.6	935.2	645.1	690.6	1404.9	459.8	1637.6	107.2	864.5	4023.3	24.9	2339.3	2364.5	16946.0	19259.5							
OTHER V.A. *3	364.4	372.6	220.6	110.4	264.5	12.1	109.3	308.0	759.0	1227.6	1376.3	763.0	356.5	93.5	108.5	1358.2	3252.7	1000.2						1720.7	
GROSS INPUT	2229.0	3707.9	896.2	7263.2	2354.3	927.5	1779.7	5709.7	1662.1	5990.0	6696.0	5803.7	6475.7	751.2	8642.8	6577.8	25817.4	31247.4							

PRIMARY INPUT

Source: 'Shetland Economy 1976/77 – Structure and Performance' by I. H. McNicoll and G. Walker.

Source: 'Shetland Economy 1976/77 – Structure and Performance' by I. H. McNicoll and G. Walker.

*1 Intermediate demand is that demand which remains within the production process i.e. demand that does not go outside of the local economy. A good example of this in the Shetland context would be the fresh fish that is used in Shetland for fish processing, i.e. a local product being used to make something else. This would be described as 'the intermediate demand for fish'. Fish caught locally and directly exported (not processed locally) would not obviously qualify for such a description.

*2 Unrequited receipts – the input-output table is fundamentally like a balanced set of accounts – someone sells something to someone else who buys it. Unrequited receipts refer to the situation where there is an imbalance for example, if someone living outside of Shetland sends some money home, but there is no sale of goods or services out of Shetland to offset this income and create a balance. This is then an 'unrequited receipt'.

*3 Other V.A. (Value Added). Other value added refers to what can be called 'gross profits'. In terms of those who have businesses, gross profits = net profits + depreciation + corporation tax (the reason corporation tax was included here was that it would not be possible to exactly calculate, as it is not always paid in the year in which it has been accrued). In terms of households, gross profits = income tax payments + savings. Another item included in 'other value added' is inventory adjustments where businesses have changes in stocks. Again, this is an item not possible to directly measure at any one time for the whole economy, so is accumulated to balance, as in this Table.

*4 Agriculture is given a negative rating here because within the industry there is a very large element of subsidy, and a balancing procedure is required as agriculture spends more than it earns.

In 1981, two major trends noted in 1976 have continued to influence the economy, namely:—

1. The change in the structure of the economy, and its dependence on the oil related construction business — especially the £1200 million Sullom Voe Project.

2. The continuing rise in the cost of living.

Meanwhile, changes have occurred within the traditional industries. The fish catching and processing industry has been badly affected, in line with the UK fishing industry as a whole, as a result of there being no agreed Common Fisheries Policy within the European Economic Community (E.E.C.) Until there is an agreed E.E.C. policy the future of both fish catching and processing remain uncertain. Even if there is established an E.E.C. fishing policy and a Fishing Plan for the Shetland area can be established, it is not envisaged that there will be much expansion in fish catching in the near future, but fish processing could expand.

Similarly, the agriculture and knitwear industries have shown no great increase in output since 1976, the future is also uncertain although a substantial expansion in part time employment could occur within the next five years.

Against this backdrop, one has, in 1981, two very fundamental changes occurring in the economy:—

A. The rundown of the construction phase of the Sullom Voe Project, on which about 1000 temporary jobs have directly depended and many others indirectly.

B. The loss of Assisted Area status for Shetland which will be effective from 1982.*

By 1982, the construction phase of the Sullom Voe Terminal is due to be completed and over 1000 local people in Shetland will have to find alternative employment from the catering (approximately one in four of all workers on the Sullom Voe Terminal Project are in catering — cooking, cleaning, etc.), labouring, and other unskilled or semi skilled work, at which they have been able to earn a good living, often by working long hours.

The traditional industries of fish catching and processing, agriculture and knitwear, to which these people might have hoped to return, are, as indicated, in an almost static state, and could probably at most absorb 100 jobs over this period, and perhaps 400 over five years.

It is therefore to other new business and industrial development that one must look to redeploy a substantial part of this labour pool, which if indirect employment is included, could amount to 15 per cent of the total adult working population.

This is likely to be particularly difficult if one considers that both new and existing business and industrial developments require investment for growth. Investment from Central UK Government funds and E.E.C. funds has been of major importance in aiding development projects in Shetland since the Oil Era began. If Shetland is to become a non-assisted area in 1982, in terms of investment aid from the UK Central Government, this will act as a double blow, viz:—

(i) it will close the doors to UK Government grants and other aid for business and industrial development in the area;

(ii) it will, worse still, close the doors to aid from the E.E.C. such as European Regional Development Funding and funding from the European Investment Bank. This is because, if the government of a member country of the E.E.C., such as Britain, does not recognise an area as worthy of 'assisted status', then the E.E.C. will not consider it to be so either.

What then, one might ask, have been the effects of investment from the UK Central Government and the E.E.C. on the Shetland economy to date? The answer can only be: very considerable.

Shetland was downgraded from Development Area to Intermediate Area status on 1st August 1980, and on present plans a further downgrading to non-assisted status will take place in August 1982.

INVESTMENT IN SHETLAND 1971-1981

1. INVESTMENT BY UK CENTRAL GOVERNMENT AND ITS ASSOCIATED BODIES

The Investment attitude of UK Central Government and its associated bodies has been fundamental to the economic development of a community such as Shetland. This investment could perhaps be categorised as follows:—

(a) Rate Support Grant;

(b) Regional Development Grant Aid;

(c) Investment through the Highlands and Islands Development Board;

(d) Investment through the Scottish Development Agency;

(e) Investment by the Industrial and Commercial Finance Corporation.

(a) **Rate Support Grant Investment**

Rate Support Grant investment has fallen off proportionally as the total rateable value of property in Shetland continues to rise. The rateable value has been particularly affected by oil related developments at Sullom Voe, and since the Terminal began its operational phase in 1978, has risen dramatically. In addition, there was a revaluation in April 1978 which reflected the tremendous rise in the cost of land and property since 1971* — one of the major factors in the rise in the cost of living (see section later on in this Chapter on 'Cost of Living' for details). Total rateable value figures for Shetland can be compared as follows:—

1970/71	£162,233
1976/77	£463,641
1978/79	£6,234,214
1979/80	£15,529,780
1980/81	£25,311,238

Not only have rates paid risen very greatly due to the values of property having altered so much from 1971 to 1978 but also, since 1978/79 the rate poundage has risen from 45p to 75p for non-domestic property (a rise of 66.6 per cent), and from 42p to 72p for domestic properties (a rise of 71.4 per cent). Even allowing for inflation, these rises have been very substantial within this context.

The most significant feature of the Rate Support Grant in recent years has been that it has remained static (for example, it was £5.164 million or 82.8 per cent of the rateable value of Shetland in 1978, compared to £6.757 million or 26.7 per cent in 1980/81), while the rates raised from Shetland ratepayers have increased enormously.

It is therefore evident that the Shetland ratepayers (amongst whom the Oil Industry is the main contributor) have funded much of the massive investment in new infrastructure which has been provided to serve the expanding population and the Oil Industry.

(b) **Regional Development Grant Aid**

A change in the Regional Policy of the UK Government was announced in July 1979, whereby Shetland was to be gradually downgraded from 1980 to 1982, when the Islands would lose their 'Assisted Area' status.

Until August 1980 Shetland was a development area which meant, amongst other things, that development projects in the Islands would attract Regional Development Grants. Grants were available, under the Industry Act 1972, for 20 per cent of the capital cost of buildings, plant and machinery for manufacturing industry, and from 1972-1980 it is known that at least £132,000 was granted to the various private business ventures in Shetland. Information was available only for payments of Regional Development Grant of £25,000 or more. This is likely to exclude the majority of cases particularly those relating to small businesses. This source of development aid has now been cut off.

1971 was the date of the previous valuation of property in Shetland, on which the rates were levied until 1978.

UK Government aid has also been available in the form of the Training Opportunities Scheme (TOPS) in Shetland. The Training Services Division (TSD) of the Manpower Services Commission has maintained three long standing courses exclusive to TOPS at the Lerwick Further Education Centre — machine knitting, copy/audio typing and basic cookery. From 1978-80, 34 people successfully completed these courses, and in addition 30 to 40 people were sponsored to leave Shetland for TOPS training on the Scottish Mainland.

The TSD has also provided direct services to industry for an economic fee and provided short training courses to employees on their firms own premises. From 1978-80, 65 employees in Shetland were thus trained in welding, fork-lift driving, slinging and lifting. Sponsored training was also given in Contractors Plant Repair and Maintenance to four Shetlanders in the year 1979/80.

Further, by the Employment Transfer Scheme many successful applications for the facilities noted below were received from unemployed workers moving to employment within Shetland viz:—

Facility	No. of Successful Applications		
	1977/78	1978/79	1979/80
Settling in grants	488	195	78
Fares	258	179	86
Household removal	17	19	8
Temporary Separation Allowance	24	20	11
Continuing Liability Allowance	5	3	—
Disturbance Allowance	196	99	50
Total (all facilities)	988	515	233

Forecasts of future trends in these figures cannot be provided as the scheme is under continuous review.

(c) **Investment through the Highlands and Islands Development Board**

The Highlands and Islands Development Board was established on November 1st, 1965, as a result of the Highlands and Islands Development (Scotland) Act of 1965. The Act charges the Board to assist the people of the Highlands and Islands to improve their economic and social conditions and to enable the Highlands and Islands to play a more effective part in the economic and social development of the nation.

The Board consists of seven members — four full time members with executive responsibility, including the Chairman and the Deputy Chairman, and three part time members. All seven are appointed by the Secretary of State for Scotland. Supporting the Board are about 250 members of staff from a wide variety of backgrounds, with the skills and expertise to service the range of activities with which the Board is involved. The Board has an office in Lerwick with two development officers, to further its activities in Shetland, which have been of considerable importance to the Community.

Unlike other Government assistance, assistance approved by the Board is entirely at the discretion of the Board, each project being considered on its own merits. In any one project, investment can be made up to £250,000 or up to £400,000 with the approval of the Scottish Office. Normally, the Board can make funding available to cover up to 50 per cent of the cost of a project, but if the project has special development value, this can go up to as high as 70 per cent. The main elements in the funding given are loans and grant assistance, but the Board can also take equity. The Board also offers assistance for social developments, the social development assistance usually going to non profit making public bodies for items such as assistance with television reception aerials, public halls, etc. — all amenities to help retain people and avert population drift from the area.*

The normal limit of grant aid for social development purposes offered by the Board is £6,000, but under very exceptional circumstances, this can go as high as £12,000.

HIGHLANDS AND ISLANDS DEVELOPMENT BOARD ASSISTANCE TO SHETLAND (1979 prices) 1971-79

Table XIII

	Grant (£'000)	Loan (£'000)	Jobs Created	Jobs Retained	No. of Projects
TOURISM					
Hotels: New	179.3	121.2	40.75	0	4
Hotels: Improvements & Extensions	202.1	203	46	25	6
Self-Catering	38.5	14.9	5.25	0	3
Bed & Breakfast	0	0	0	0	0
Other Accommodation	39.7	15.2	14.75	0	9
Staff Quarters	0	0	0	0	0
Catering	68.0	38.6	27.25	0	6
Marine Recreation	3.1	6.4	1.25	0	1
Other Recreation and Tourist Amenities	11.9	0	3	0	1
TOTAL	542.6	399.3	138.25	25	32
INDUSTRIAL DEVELOPMENT AND MARKETING					
Mining & Quarrying	18.2	78.7	12	0	3
Fish Processing	197.2	745.0	105.5	223	15
Boatyards & Marine Engineering	0	42.8	14.5	0	2
Crafts	13.3	9.7	23	0	8
Printing	1.4	40.9	16.5	0	4
Other Manufacturing	41.3	79.3	21	61	7
Building Industry	14.6	119.5	20	1	7
Service Industries	160.8	138.3	71.5	10	29
Transport	104.8	15.6	7	1	7
Others	2.0	0	1	0	1
TOTAL	553.6	1269.8	292	296	83

FISHERIES

Fishing Boats: New	23.6	122.3	9	0	7
Fishing Boats: Secondhand	82	1888.4	133	49	67
Services	0	10.2	2	0	1
Fish Farming	39.7	0	4	0	1
Fresh Water Fisheries	0	0	0	0	0
TOTAL	145.3	2020.9	148	49	76

LAND

Farm Development					
general	16.6	13.6	3.5	2	4
sheep	0	0	0	0	0
beef	1	0.4	0	2.5	2
dairy	5.7	87.6	7	5	6
crops	0	0	0	0	0
Horticulture Development	1.1	0.6	0	1	1
Machinery Syndicates	29.1	14.2	0	9.5	22
Contracting Services	2	2.4	2	0	2
Services to agriculture, horticulture, forestry, not elsewhere specified	17.6	0	1.5	0	3
Slaughterhouse processing	22.7	74	6	0	2
Other	0	14.3	1.5	0	1
TOTAL	95.8	207.1	21.5	20	43
Publishing	0	2.1	1	0	1
OVERALL TOTAL	1337.3	3899.2	599.75	390	235

Many business advisory services are offered by the Board, for example, in the fields of accountancy, marketing and production. The current (1980/81) Board spending is about £25 million, of which £20 million comes directly from the UK Central Government through the Scottish Office; the £5 million balance is from repayment of capital and loans, and rents from factories and other buildings. Of this £25 million, about £2 million will be spent on grants and loans schemes; £7 million on advance and custom built factories for leasing, and on hotels and tourist information offices; and £1.7 million on the promotion of industry and tourism, and research and survey work. The remainder is spent on the salaries of administrative staff.

A summary of the Board Assistance to Shetland from 1971/79, in terms of 1979 monetary values is seen on the accompanying chart. This shows that in these years the work of the Board in Shetland created approximately 600 jobs and retained 390; loans of £3.899 million were made, and grants of £1.337 million (at 1979 monetary values). The fish catching and processing industry was perhaps the biggest beneficiary, but Board assistance covered all aspects of the economy.

(d) **Investment through the Scottish Development Agency**

The Scottish Development Agency (SDA) which was established in December 1975, is funded by the UK Central Government. The Chairman of the SDA, like the Chairman of the Highlands and Islands Development Board, is appointed by the Secretary of State for Scotland. The SDA has powers to provide a wide range of assistance to industry including the provision of factories, advice on management and marketing, and the provision of equity or loan capital. The Agency's remit extends throughout Scotland, but in the Highlands and Islands it does not involve itself in activities which fall within the range of assistance offered by the Highlands and Islands Development Board (HIDB).

For example, the HIDB carries out the functions of the SDA in other parts of Scotland by offering financial assistance, and manufacturing and technical advisory services to small businesses.

The HIDB is also responsible for factory building and letting, but has available technical support from the SDA.

However, the SDA operates independently in other fields, particularly in its powers to invest in companies through loans and equity, and in land renewal work which has cleared industrial and war time dereliction throughout the Highlands and Islands.

To date, the SDA has been active in one major project in Shetland, clearing and reclaiming land for new playing fields at Lerwick. This project which began in February 1980, has also involved building access roads and car parking. By the time the necessary works have been completed in 1982, it is estimated that the SDA will probably have spent in excess of £600,000.* This scheme is of considerable value in creating central sports facilities which have been very much needed by the increasing population.

(e) **Investment by the Industrial and Commercial Financial Corporation (I.C.F.C.)**

Backed by the Bank of England** and all the UK high street banks, ICFC was formed in 1945 to help regenerate British Industry following the war effort. By 31st March 1980, ICFC had £378 million invested in over 3,300 separate businesses. As a provider of long term loan and equity finance, as well as hire purchase and leasing facilities for small to medium sized businesses, it is recognised as the largest organisation of its kind in the world. Despite its size, the smallest investment level is still £5,000 and in a typical year about 50 per cent of customers will require amounts of less than £50,000. Two areas in which ICFC has been particularly active in recent years have been in helping businesses get started and in assisting managers buy control of companies they manage.

Since 1975*** ICFC investment in Shetland has exceeded £2 million (based on December 1979 money values) and 20 per cent of this has been in assisting companies start up.

The Shetland Islands Council also contributed £250,000 to this scheme.
**Hence its central UK Government connections.*
****ICFC was not notably active in Shetland before 1975.*

A breakdown of ICFC's investment in Shetland is as follows:—

Industry	Amount (£)	%
Oil Related	900,000	42
Building & Building Supplies	850,000	40
Property	200,000	9
Fish Processing	100,000	5
Retail	50,000	2
Hotels	50,000	2
	2,150,000	100

Having briefly summarised investment by the UK Government and its associated bodies since the Oil Era, it is relevant now to consider the investment made in Shetland by the major E.E.C. bodies.

2 E.E.C. INVESTMENT

There have been two main sources of E.E.C. investment in Shetland to aid the economy to cope with oil related developments:—

(a) The European Investment Bank

(b) The European Regional Development Fund

(a) **The European Investment Bank**

The European Investment Bank is the European Economic Community's (E.E.C.) banking institution, and long-term loans for industry and infrastructure have been available from it since Britain's accession to the E.E.C. on 1st January 1973. Since then over £1650 million has been provided in long term loans for industry and infrastructure in the United Kingdom, of which £65.65 million had been invested in the Shetland Islands by the end of 1980 (this sum does not include £45.1 million in loans which were made for the development of the Thistle, Beryl and Frigg fields).*

The European Investment Bank borrows funds on the world's capital markets, and working on a non-profit making basis, lends or gives its guarantee for loans granted by other bodies towards the capital cost of projects which promote a balanced and steady development of the community by:

(i) helping to stimulate the development of less favoured regions, including all UK Assisted Areas;

(ii) modernising or converting industries or creating new activities, to offset structural difficulties affecting certain sectors;

(iii) serving a 'common interest' of several Member States on the Community as a whole (e.g. reducing dependence on imported energy sources or improving communications between Member Countries).

The major loans which the European Investment Bank can make are for not more than half the cost of the fixed assets needed for a project. Such loans are for generally not less than £2.5 million over a term normally seven to 12 years for industrial projects, and up to 20 years for infrastructure. Repayment is generally made by equal half yearly payments of capital and interest after a franchise period of up to five years, depending on the project completion date. Loans are usually in several currencies and the Bank's interest rates closely follow its borrowing costs and consequently vary from time to time with money market rates. However, the interest rates are fixed for the duration of the loan.

---

*See Chart 'European Investment Bank loans in Shetland 1973-1980.

EUROPEAN INVESTMENT BANK LOANS IN SHETLAND 1973-1980　　　　　　　　　　　　　　　　**Chart XXVIII**

Date	Amount of Loan (£ million)	Project	Comments
October 1980	5	Providing two 8.1 megawatt diesel generators and two 5.2 megawatt gas turbine units at the North of Scotland Hydro Electric Board's Lerwick Power Station.	Loan for 15 years at 11.5% interest to help meet the cost of increasing substantially electricity generating capacity in Shetland in response to demand which has built up rapidly with growth of oil related activity.
August 1980	5	Help with infrastructure investment to Shetland Islands Council for road construction, housing, water and sewerage provision.	Loan for 10 years at 11.25% interest to serve the development of the North Sea oil industry and also the Islands traditional activities of agriculture, fishing and knitwear. Total project cost: about £42 million.
January 1980	6.23	Towards construction of fourth crude oil jetty, cargo jetty, tug harbour and associated facilities by Shetland Islands Council.	Loan for 10 years at 11.55% interest. Sullom Voe oil port benefits from regional development and a total of 650 jobs should be created by these activities in an area where employment opportunities were restricted.
December 1979	11.07	Towards construction by Shetland Islands Council of fourth crude oil jetty, cargo jetty, tug harbour, and range of associated facilities at Sullom Voe Port.	Loan for 10 years at 11.55% interest.
August 1978	10.75	Extension of Sumburgh Airport by the Civil Aviation Authority. New Terminal building, helicopter landing strip, parking, etc.	Loan for 10 years at 8.9% interest covers half the estimated cost of the new developments at Sumburgh.
December 1976	10.7	Ninian Pipeline System which was constructed by BP.	Loans for 8 years at 8½% interest. Estimated cost of the System around £200 million. Pipeline system expected to handle 18% of UK's requirements or 3% of EEC's consumption by the 1980's.
December 1975	16.9	Oil tanker harbour construction at Sullom Voe by Shetland Islands Council including three T-shaped jetties, cargo jetty, tug harbour construction village, houses, school.	Loan for 10 years at 9½% interest. Project was then expected to cost about £50 million. This was when Sullom Voe Terminal facility costs were quoted at £400 million as opposed to 1981 costs of £1200 million.

Total Loans 1973-80: £65.65 million

(b) **European Regional Development Fund (ERDF)**

Shetland has also had assistance from the European Regional Development Fund since 1973 but nothing like on the scale of the massive European Investment Bank assistance. There has, for example, been some assistance given with road building, and also on a larger scale, assistance in the development of Lerwick Harbour which has amounted to approximately £1.9 million, this included £868,000 to help build the new ro-ro berth (see Chapter Three), approximately £990,000 for new quays and £23,000 towards a new 40 tonne crane.

In total, monies obtained from ERDF to the end of 1980 amounted to £2.4 million for private business ventures and £6.65 million for public sector projects, such as the Lerwick Harbour one above mentioned.

Most of the sources of assistance from the UK Central Government and the EEC which has been available to help the people of Shetland through the Oil Era are going to be closed to Shetland from 1982 if the Islands lose their Assisted Area status, as is currently planned.

This will have very serious economic ramifications, particularly at a time when assistance for new developments and new projects will be needed most, with a potential 15 per cent of the workforce unemployed.*

A review has been commissioned by the Secretary of State for Scotland of circumstances in areas such as Shetland, scheduled to be down graded to non-assisted area status in 1982. This review is being carried out with the co-operation of the Scottish Development Agency, the Highlands and Islands Development Board and the local authorities concerned.

The consultants concerned — Mr Stuart McDowall of St. Andrews University and Dr Hugh Begg of Dundee University are due to report to the Secretary of State in March/April 1981.

A further factor affecting the economy, previously touched upon, but not expanded, is the rise in the cost of living which has occurred in Shetland since the Oil Era arrived.

The Cost of Living

The Cost of Living could be considered in two categories:—

(a) The cost of Land and Housing;

(b) Other costs

(a) **The Cost of Land and Housing**

Land and property prices in Shetland through the Oil Era have depended very much on location. One could perhaps take the three areas most affected by oil developments — Sullom Voe, Lerwick and Sumburgh, and consider what has happened in each.

(i) **Land in the Sullom Voe Area**

Land in the Sullom Voe area was required for several purposes — for pipelines, the Terminal and Port developments, for Scatsta Airport, for the construction villages, and for housing.

Land for the pipelines — mainly wayleaves to allow pipelines to pass through the land — came under the terms of the Pipelines Act of 1962.

By this Act, landowners cannot prevent a pipeline going through their land unless they have a valid objection under the Act. The constructors of the pipeline can take a servitude, and compensate the landowner/crofter for the disturbance caused, and the damage done to the land. If the landowners have no valid objections under the Terms of the Act, the land can be compulsorily acquired. This never happened. The crofters concerned were all able to agree on the compensation available.

This will include some married women who have been working at Sullom Voe, and who may not have been working in full time employment before the Oil Era.

In the early 1970's the land on which the Sullom Voe Terminal and Port is now built was valued at £20 per acre for rough grazing and £200 per acre for arable. What occurred there was unique in world wide terms in that there was to be a once and for all increase in the value of the land. A thousand acres of land which might previously have sold for £20,000 to £50,000 (maximum) fetched £2 million eventually. Certain individuals and companies reasoned that if developments were going to occur it was wise to buy land in advance. This by its very nature was a speculative action and led to the local authority seeking and obtaining special powers to bring this key land under public ownership.* Some land actually had decreased in value in 1980 from its estimated value in 1977 because developments anticipated there did not materialise.

(ii) **Land at Lerwick**

A more established and easier to manage situation arose at Lerwick, in terms of land for both industrial and domestic purposes.

Lerwick Harbour Trust bought Gremista Farm land, of approximately 1500 acres, for £20,000 in 1972. From 1972, and the announcement of oil discoveries off Shetland, prices soared. Land valuation in 1981 still depends on the location and servicing available, but prime land at Lerwick Harbour is now worth from £60,000 to £150,000 per acre, if it is alongside the water front where a jetty might be developed.** At £150,000 per acre, it would probably reflect the cost of reclaiming.

Housing costs in Lerwick from 1971 to 1980 have risen (in terms of straight monetary increases) from 200 to 918 per cent. This is considering a range of properties bought in 1969/72 and resold 1978/80. An increase of around 800 per cent would be probably a fair average figure over the period. If one was to compare the Lerwick situation with the situation elsewhere in Scotland, there would still be a higher increase in Lerwick. For example, an average semi detached house in the Bridge of Allan, Central Scotland area, would have cost £5000 in 1971 and in 1981 £28,000. This would be an increase of approximately 600 per cent. In Lerwick for the same situation one could look to an 800 per cent increase. House plots in Lerwick have undergone similar increases. A serviced house plot in 1971 would cost £400, in 1980 there were plots for sale at £10,000 each.

(iii) **Land in the Sumburgh Area**

In the Sumburgh area, land in 1971 was £20 per acre for agricultural use and £100 for a house plot (usually an acre). In 1980 it was £6000 per acre for a house. The standard feuing rate was applied so that it would cost £6000 per acre whether one bought one acre or ten.

By the Crofting Reform (Scotland) Act of 1976, if land is decrofted for housing or other development purposes, half the development value goes to the landlord, and half to the tenant crofter and this is what has been happening in the Sumburgh area. This happened voluntarily before 1976 and speeded up the release of land for housing.

At the airport, land for runways does not have a value as such. Land values are based on the benefit accruing to the occupants, and what can be made of the land in terms of commercial and industrial development. Airport values are therefore artificial and liable to arise from what they have been displacing — housing or rough grazing. Again, like the Sullom Voe Terminal area, this was a once and for all event.

The great pressure on accommodation was not a one off event, and was to last until about 1979. There is still housing pressure on the local authority, but in the Lerwick rather than the Sumburgh area, where there has recently been a reduction in job opportunities.

Few young couples starting off married life can afford to build a house. Costs are such that it can be up to 50 per cent dearer to build a new house in Shetland than on mainland Britain. Most new houses are built using concrete blocks, locally manufactured, in order to save money. Concrete blocks are not the ideal material for Shetland as they retain moisture, but the cost of importing bricks is prohibitive.

See Chapter Two.

**It is only some of the land at Lerwick Harbour which has this high valuation. There is much of the former Gremista farm land which is rough hill grazing, and not of prime industrial or commercial value.*

Labour costs have also been very high since the Oil Era. Until recently there was a tremendous shortage of skilled tradesmen and wages rose to keep pace with oil related work.

Other aspects of the Cost of Living

Other aspects of the cost of living fall mainly into the categories of food, consumer durables, electricity, coal and gas and transport. A Rural Scotland Price Survey is now in being, conducted by the Aberdeen University Institute for the Study of Sparsely Populated Areas. The Autumn 1980 Report of this Survey showed quite clearly that Shetland has some of the most expensive locations in Scotland. Apart from Lerwick, prices in all locations covered in Shetland were consistently in the range of 12 to 20 per cent above Aberdeen. Passenger and freight transport costs were isolated as the main contributory factors. A reduction in transport costs to the UK Mainland would reduce overall price levels in Shetland by a significant percentage. It seems likely that some of the increased cost of living will disappear as the local economy deflates.

The State of the Economy in 1981 and the Future

What then is to be considered the state of the Shetland economy in 1981, and what of the future?

Currently three problems exist:—

(1) Up to 15 per cent of the workforce — over 1000 people — are likely to be redundant within a year;

(2) Just at the very time when Shetland most needs Assisted Area status, to help business development and redeploy these people the Islands are programmed to lose this;

(3) The cost of living/building/development in Shetland is high, because it is a remote area, and as such needs to be treated as a special case.

Oil related construction activites have exacerbated these problems to some extent, by raising wage levels and prices, and by attracting in people who have made Shetland their home.

At the same time a balance must be struck, and it is perhaps of use to consider what areas of the economy, apart from oil related construction activities, have been developing, and unlike oil related construction activities, might continue to develop apace, given the right incentives. A brief consideration of the banking sector will perhaps give some indication of how the economy has grown, and in what areas.

THE BANKING SECTOR

The banks in Shetland have all reported a considerable increase in their business over the decade 1971-1981, with an increase in the number of accounts, deposits and advances being made varying from 150 to 870 per cent in certain business areas. All felt that since the discovery of oil offshore Shetland business had increased, but that it was not possible to quantify what percentage of this increase was attributable to oil.

The banks also felt that over the past decade the banking business itself has been revolutionised. Since 1970 many more people have been persuaded to use cheque books, and there has been the development of personal credit card facilities.

Employers, partly for reasons of security, tend now to make direct wage payments into the banks and this factor has encouraged people to maintain current accounts. The banks have also increased business by offering new services such as insurance broking and loan facilities.

Within this framework, business has been affected by oil activity and each of the three major clearing banks in Shetland — the Bank of Scotland, the Clydesdale Bank, and the Royal Bank of Scotland, has more than doubled its staff since 1971.

It is not, however, the move of multi-national companies into Shetland that has increased banking business. The large companies normally have centralised banking services elsewhere and their local Shetland banking accounts are not of consequence when compared to their main banking facilities.

The business that the banks have been more concerned with has been locally based spinoff — for example, extending facilities to new businesses and developing established businesses.

The business areas where significant increases in turnover, have been noticed are these connected with the supply of food in any form (wholesale, retail, manufacturing); radio and TV dealers (there was a boom when colour TV came to Shetland in 1977 which was probably encouraged by oil related money in the economy); light engineering, hotels, and the commercial sector — for example, selling consumer goods such as newspapers to the construction villages.

The banks concluded that very few businesses could not be said to have benefitted from the economic climate and the increase in population resulting from oil. It was felt that banking in Shetland probably reflected banking in the other oil related areas, such as Aberdeen, where there has been more opportunity for business to increase.

The banks have had to bring in staff from the Scottish Mainland as there has been such a shortage of labour in Shetland. The situation changed in 1980, and they are now able, for the first time in seven years, in some cases, to have a choice of employees to recruit from locally. At the end of 1980, the banks were employing a total of 92 people.

General Comments on Oil Related Spinoff Activity

There has been such a variety of different occurrences within the business community in Shetland that it is perhaps iniquitous to select a few examples or 'case histories' to illustrate what have been areas of business growth directly attributable to oil developments, but operating on the principle that case histories can sometimes illustrate more clearly what is happening than a thousand dry statistics, the following case histories have been chosen:-

(1) **J. & M. Shearer** — a business based on the herring fishing which was bought over, and diversified with the Oil Era;

(2) **John Leask & Son** — Travel agents, a Shetland family owned business which expanded with the Oil Era;

(3) **J. M. Ironside** — a food wholesaling business started as a result of the opportunities created by the Oil Era;

(4) **Hay & Company Limited** — see Chapter Three;

(5) **Malakoff and Moores Boatyards** — traditional Shetland boatyards serving the fishing industry; bought over, and since 1974 managed by a joint venture between the Wood Group of Aberdeen and Lithgows of Glasgow. This business has diversified and expanded with the Oil Era;

(6) **Bolts** — Shetland owned company based on car hire and garage maintenance work. This business expanded, diversified, and is now exporting experience gained in Shetland by establishing a branch in Mossmorran, Fife;

(7) **Shet-Link** — a road haulage business begun by local people and expanded with oil related business;

(8) **Miller Construction Northern Limited** — an example of the scale of the construction work which occurred in Shetland (for the detailed case histories of these businesses, see the Appendix to this Chapter).

Having briefly illustrated past developments by the case histories it is appropriate to look to the future. For the future, it is probable that the structure of the Shetland Economy will again undergo a change within the next five years and that the Oil Terminal and Port at Sullom Voe will supplant the oil related construction activities in importance as a contributor to the economy, followed closely by, if not on a par with, marine supply services. The fishing industry will hopefully be up there still, as one of the mainstays of the Shetland economy, alongside Shetland's newest 'traditional' industry — the Oil Industry.

It is to servicing the oil industry that the economy will probably have to look to redeploy many of those made redundant with the rundown of the construction phase at Sullom Voe. It is to be hoped that this might be achieved without an uncomfortable hiatus and that Shetland might be able to maintain Assisted Area Status. Further, the evaluation of the freight subsidy situation currently being undertaken by the Scottish Office, could do much to help the cost of living situation in Shetland and make exports competitive. The combination of Assisted Area Status with an equitable cost of living situation, plus Shetland's unique location an accessibility to the oilfields, could augur well for the future economic wellbeing so necessary to the survival of this community.

Table XIV

EMPLOYEES IN EMPLOYMENT IN SHETLAND BY BROAD INDUSTRY GROUP, 1958-77

Year	Primary	Fish Processing	Knitwear	Total Manufacturing	Construction	Services	Total
1958	623	365	82	878	996	2557	5054
1959	759	343	96	687	1188	2770	5404
1960	667	384	80	707	734	3025	5133
1961	566	447	98	750	669	3215	5300
1962	629	571	103	874	756	3158	5417
1963	686	598	95	880	746	3193	5505
1964	674	512	102	776	693	3264	5407
1965	543	546	126	832	520	3131	5026
1966	591	539	225	938	539	3220	5288
1967	555	559	216	912	451	3170	5088
1968	665	509	218	973	539	3064	5241
* 1969A	755	451	333	1102	558	2852	5267
* 1969B	755	451	333	1068	558	2886	5267
1970	749	505	401	1109	510	2853	5221
** 1971A	625	808	362	1426	620	2849	5520
** 1971B	279	730	314	1302	489	2915	4985
1972	325	685	344	1282	627	3009	5243
1973	283	533	477	1287	719	3066	5355
1974	304	547	389	1193	919	3379	5795
1975	340	288	271	801	1396	3429	5966
1976	316	544	144	985	1503	4193	6997
1977	453	627	112	1123	2164	4829	8569

*The Standard Industrial Classification was revised in 1968. In 1969 statistics were compiled on the basis of both the 1958 and 1968 SIC (1969A and 1969B figures respectively) for comparison.

**In 1971, the method of collecting employment statistics changed from the National Insurance Card Count to the Annual Census of Employment. Both sets of figures are included in the table (1971A and 1971B respectively) for comparison purposes. The marked change in the primary sector is partly due to the different handling of share fishermen in the two methods of counting.

Source: 1958-71 National Insurance Card Count
1971-77 Annual Census of Employment

Chapter Ten

Summary

This book has been based on facts. The facts relate to how Britain's most northerly community, Shetland, became embroiled with the international oil industry, and how since the 'Oil Era' began in 1971, great changes have occurred in this remote and beautiful area.

These changes have resulted in a tremendous investment being made, not only in terms of financial provision, but also in terms of human endeavour, to create the facilities needed to help Britain withstand the energy crisis, which has become a feature of the latter part of the twentieth century.

An investment of approximately £1200 million has been made by the oil industry at Sullom Voe; an investment of approximately £150 million has been made by the Shetland Islands Council to provide the Port of Sullom Voe, infrastructure and all the necessary facilities to allow the Shetland community to cope with these major developments; investments have been made by the UK Government and its associated organisations, such as the Civil Aviation Authority, the Highlands and Islands Development Board and the Scottish Development Agency. Help has also most generously come from the European Economic Community in the form of loans from the European Investment Bank and European Regional Development grants. Many private individuals and companies have also made considerable investments within their own spheres of activity.

In terms of human endeavour, literally millions of manhours have been expended to achieve the development of the facilities that are now available in Shetland.

These facts are perhaps stranger than fiction, for who would have anticipated such radical changes occurring in this remote island community even ten years ago?

The local Shetland community has not remained untouched by these events. Shetland is a society of individuals with its own way of life. Economic change is never without social consequence, and the community has approached the past decade and its extraordinary events with a sense of realism and forethought in its efforts to try and preserve the best of the traditional way of life, whilst at the same time incorporating change. It is to be hoped that the achievements of the past ten years can be consolidated as Shetland's Oil Era enters its second decade.

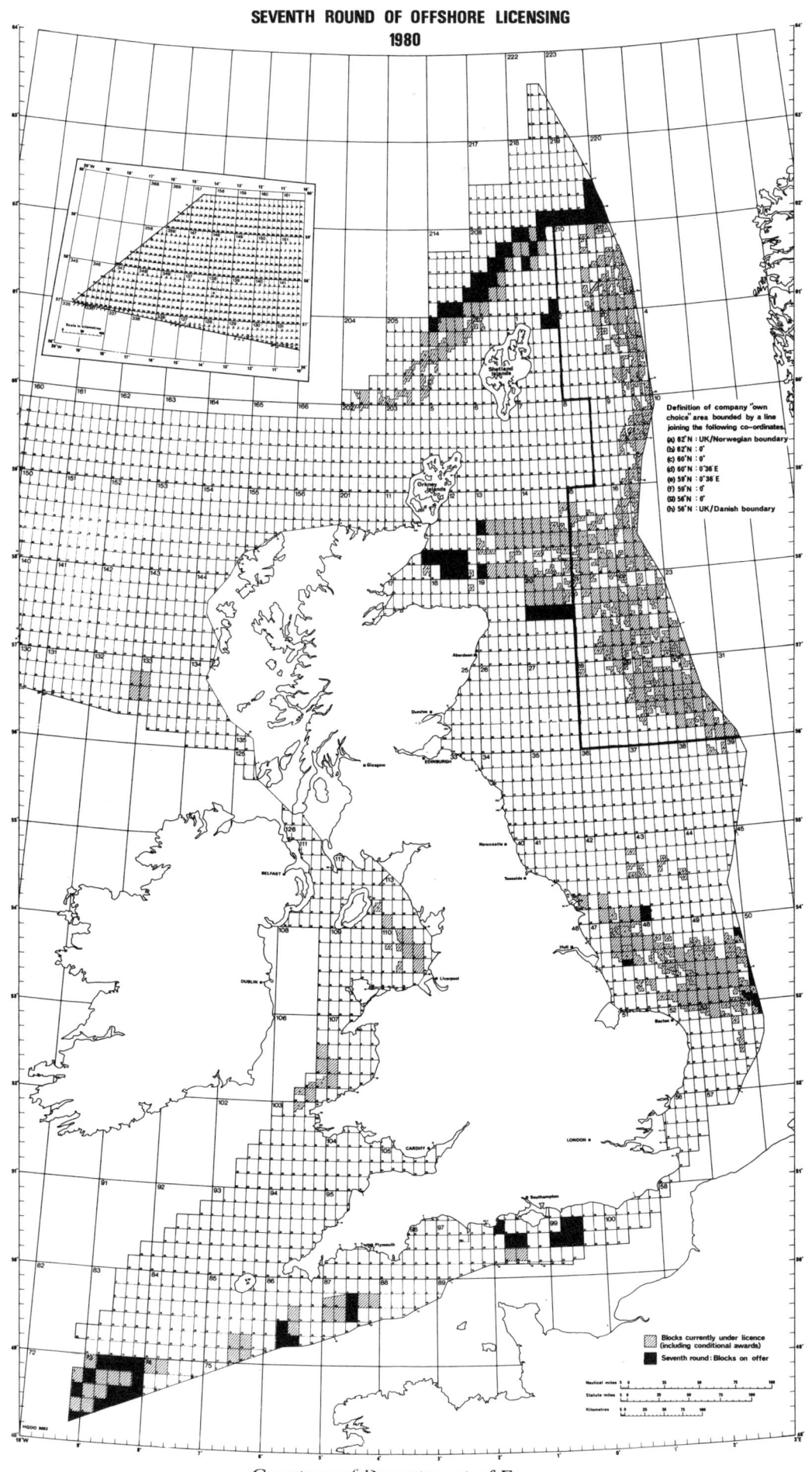

Courtesy of Department of Energy

APPENDIX TO CHAPTER ONE

EXPLORATION FOR AND DEVELOPMENT OF OIL AND GAS FIELDS

LICENCE BLOCKS NORTH OF 59 DEGREES NORTH
(i.e. within the Shetland area)

LICENCE BLOCKS

For licensing purposes the UK Continental Shelf is divided into rectangular areas or quadrants bounded by lines of latitude and longitude. Every quadrant is divided into 30 blocks, each block measuring 12 minutes of longitude by 10 minutes of latitude. The blocks are numbered from 1 to 30 within each division, and each block constitutes an individual concession.

For example: 211/18 = Quadrant 211, Block 18

The size of the blocks decreases as the lines of longitude converge towards the North. Thus, UK sector blocks vary in size from 215 to 250 square kilometres. At the present time, (1980) the area of the designated UK Continental Shelf totals some 642,806 square kilometres, of which about 68,913 square kilometres have been allocated production licences by the UK Government Department of Energy.

The area to which production licences have been allocated expands with each "Round" of licensing.

There have now been seven "Rounds" or allocations of offshore production licences on the UK Continental Shelf — in 1964, 1965, 1970, 1971/72, 1976/77, 1978/79, and in 1980. Of these seven Rounds, only the first, in 1964, did not involve the area affecting Shetland, North of 59 degrees North.

Terms and Conditions of Offshore Licensing

There are two types of offshore licence issued by the UK Government Department of Energy;
(i) Exploration Licences
(ii) Production Licences

(i) **An Exploration Licence** permits only the conduct of geological and geophysical surveys and the drilling of shallow wells to explore for oil and gas. It does not allow the extraction of oil and gas. It is non-exclusive and permits the exploration of any designated part of the UK Continental Shelf except areas already covered by production licences.

Confidentiality of exploration information is important for both commercial and political reasons. A condition of exploration licences is that well samples must be kept for the duration of the confidentiality period (now five years). Within this confidentiality period the Secretary of State for Energy is empowered to require that any part of a sample be surrendered to him (this condition also applies to production licences).

After the expiry of the period of confidentiality the samples must be offered to the Department of Energy for preservation. If within six months the Department indicates that it does not wish to acquire the well samples, they may be preserved or disposed of as the licensee wishes.

Prior to the 5th Round, exploration licences ran for three years and cost £1000 per annum. Since the 5th Round, the duration of exploration licences can be extended to six years on request and the rent is increased to £2000 per annum, plus a £500 application fee.

(ii) **A Production Licence** gives the licensee the exclusive right to search for and get oil and gas in a specified area.

Areas under licence can be held for up to two terms:

The initial term of the licence is six years, after which licensees can opt (subject to certain conditions) for a 'continuing term' of 30 years, in respect of 50% of the area originally licensed. The concept of a minimum retainable area is introduced, i.e. 30 sections (there are 120 sections in each block) except that where the originally licensed area was less than 30 sections, mandatory surrender of part of the area will not be required.*

Government Participation

Major changes were introduced into the management of the UK Continental Shelf, when, with the 5th Round of licensing in 1976/77, the Government decided to participate directly in the exploitation of the oil and gas resources. The method of participation chosen was for either of the two state owned companies, the British National Oil Corporation (BNOC) or the British Gas Corporation (BGC) to receive a majority 51% interest in all licence blocks, paying their share of expenses as they were incurred. BNOC and BGC were also entitled to special allocation of blocks outside of official licensing rounds. The 5th Round and the 6th Round licences were awarded subject to two conditions.

1. Agreement between the Department of Energy and the potential Licensee (including BNOC) on a compulsory exploration programme for the blocks to be licensed.

2. The conclusion of a Joint Operating Agreement by BNOC and each group of co-licensees satisfactory to the Secretary of State for Energy.

The Present UK Government has instituted changes in the role of BNOC, and thus its UK Continental Shelf Management. For the 7th Round of licensing the situation is thus:

Production licences are granted on the basis that BNOC will have the option to take at market value up to 51% of the petroleum produced in the licensed area, except where BNOC holds an equity interest in the Licence, in which case BNOC's option will relate to each of its co-licensee share of production.

In addition to the above mentioned petroleum, BNOC has considerable oil trading activities and sells the Government's as well as its own royalty crude. The result of this was in 1980, BNOC was selling over 500,000 barrels of oil per day on world markets.

The BNOC has however, no longer a statutory role as advisor to the government and no longer sits on operating committees of licences in which it has no equity stake. Furthermore it has now to pay Petroleum Revenue Tax (PRT) like any other licensee and has to finance any developments it is involved in outwith the National Oil Account, which it previously had access to.

The cost of production licences. At the time of the 7th Round the following applied:

a. In respect of the first term of the licence a non-recurrent fee of £250 for each square kilometre comprised in the licensed area, was payable, for blocks in the area which the Government chose to licence.

b. For blocks within an area where the Government allowed the oil industry to select acreage, a non recurrent fee of £5 million for each block so licensed was payable on the grant of the licence.

c. In respect of the licensee requiring the licence to continue after the expiry of the first term, the sum of £300 for each square kilometre in the continuing area of the licence is payable for the first year; for the second year £600 per square kilometre and similarly by annual increments until an annual sum of £4500 will be payable for each square kilometre within the area licensed.

d. A royalty at the rate of 12½% either in kind (crude petroleum) or by value on all quantities of petroleum won and saved, is also payable to the Government.

Production Licences are issued solely at the discretion of the Secretary of State for Energy and criteria for consideration in issuing licences include: technical competence and ability to finance and undertake a programme of exploration and produciton, past performance, if applicable; contribution to the UK economy; Agreement to allow trades union representatives reasonable access to the workforce, and training facilities available to the workforce.

For further information see Petroleum (Production) Regulations, 1976, Statutory Instruments 1976 no. 1129; 1978 amendments in Statutory Instruments 1978 no. 929; and 1980 amendments in Statutory Instruments 1980 no. 721, which came into operation 14th June, 1980.

Summary of the Licensing Situation Offshore Shetland to November 1980

North of 59 degrees North there are 27 companies* who act as 'operators' or administrating companies for the total of 70 licences currently valid, covering 140 blocks or part blocks. Each licence may be held by one or several companies. It is normal for there to be only one operator per block licensed. In the case of a group of companies holding a licence concession one of their number is usually elected to act as operator.

A total of 188 blocks have been licensed. Since 1965, 48 of these blocks have been relinquished, leaving a total of 140 blocks currently licensed in this area.

Round	Date	No. of Licences Issued	No. of Blocks Licensed
2	24/11/65	2	13
3	6/6/70	6	17
4	16/3/72	58	131
5	9/2/77	6	8
6	3/8/77	8	12**
7	11/8/80	N/A	N/A

BLOCKS OFFSHORE SHETLAND (NORTH OF 59°N) OFFERED BY THE DEPARTMENT OF ENERGY IN THE SEVENTH ROUND OF LICENSING IN MAY 1980

Summary

Blocks offered fell into two categories:

1. Those selected by the Department of Energy, numbering in total 34, namely:

    1/4, 1/5, 206/1, 208/9, 208/10, 208/13, 208/14, 208/16, 208/17, 208/18, 208/21, 208/22, 209/1, 209/2, 209/4, 209/8, 209/12, 209/30, 214/25, 214/27, 214/28, 214/29, 218/29, 218/30, 219/20, 219/25, 219/26, 219/27, 219/28, 219/29, 219/30, (220/16 + 220/21 + 220/22),*** 220/26, 220/27.

2. An area bounded by a line joining the following co-ordinates
    a. 62° North: UK/Norwegian boundary
    b. 62° North: 0°
    c. 60° North: 0°
    d. 60° North: 0° 36′ East
    e. 59° North: 0° 36′ East
    f. 59° North: 0°

    in which companies can select the acreage for which they wish to have production licences.

See chart of licencees for further details.
**13 blocks were offered, one, 208/20, was not taken up by any licensee.*
***220/16, 220/21 and 220/22 are treated as one block.*

UK LICENCE BLOCKS NORTH OF 59° NORTH* (as at end January 1981)

Operator	Licences	Round	Block Licensed Originally	Blocks Currently Licensed
Amoco (UK) Petroleum	P184	4	210/24, 211/27, 3/11, 9/3, 9/27	210/24a, 211/27, 3/11a, 9/27a
	P235	4	9/4 (assigned from Numac 15.3.78)	9/4a
	P320	6	206/3, 208/19, 208/24	206/3, 208/19, 208/24
Arco Oil Production Inc.	P164	4	205/26	205/26a
	P188	4	210/30, 211/11	210/30a, 211/11a
BP Petroleum Development Ltd.	P165	4	205/22, 206/8	206/8
	P166	4	202/3, 204/27, 204/28	202/3a, 204/27a, 204/28
	P193	4	211/7, 211/12	211/7a, 211/12a
	P194	4	9/5	9/5a
	P198	4	3/29	3/29a
	P199	4	3/8	3/8a (Chevron is operator for part which is Ninian Field)
	P206	4	211/2	211/2a
	P268	5	3/10b	3/10b
	P319	6	208/15	208/15
	P329	7	3/8b	3/8b
	P338	7	9/24b	9/24b
	P365	7	210/15b	210/15b
	P366	7	211/11b	211/11b
British Gas Corporation	P281	5	3/9b	3/9b
	P303	6	214/30	214/30

Further licences in the Seventh Round will be issued later in 1981.

Operator	Licences	Round	Block Licensed Originally	Blocks Currently Licensed
*The British National Oil Corporation	P290*		205/10, 206/6, 208/27, 209/9	205/10, 206/6, 208/27, 209/9
	P202	4	3/3	3/3 (Chevron is operator for Ninian Field)
	P236	4	211/18	211/18a
	P268	5	3/24b	3/24b
	P277§	5	9/14b	9/14b
	P306	6	208/26	208/26
British Sun Oil Co.	P200	4	211/22	211/22a
	P325	7	2/4	2/4
Chevron Petroleum Limited	P203	4	3/7	3/7a
	P202	4	3/3	3/3
	P199	4	3/8	Part 3/8a
	P190	4	3/23a	3/23a
	P234	4	2/10a, 3/28a	2/10a, 3/28a
	P326	7	2/10b	2/10b
	P327	7	2/15	2/15
	P328	7	3/7b	3/7b
Conoco Limited	P103	3	9/18, 9/19	9/18a, 9/19
	P104§	3	3/5, 211/19, 211/24, 211/25	211/19a, 211/24a, 211/24b, 211/25a
	P204	4	210/3, 211/28, 3/2, 9/21	211/28a, 3/2
	P293	5	211/24c, 211/25b	211/24c, 211/25b
	P334	7	9/3	9/3
	P336	7	9/12b	9/12b
Elf Oil Exploration & Production (UK) Limited	P168	4	205/20, 206/7, 206/11, 206/16	205/20a, 206/7a, 206/11a, 206/16a

FOOTNOTES: * This was a sole licence situation where the award of a licence can take place outside of a round to a state owned company, such as the British Gas Corporation or the British National Oil Corporation.

§ It is possible for a licence to be shared by two companies which may be operators of different blocks within that licence.

Operator	Licences	Round	Block Licensed Originally	Blocks Currently Licensed
Esso Petroleum Co. Limited	P169	4	202/2, 205/19, 206/12, 206/13	202/2a, 205/19a, 206/12, 206/13a
Gulf UK Offshore Exploration Co. Limited	P301	6	209/6	209/6
Hamilton Brothers Oil Co. (GB) Ltd.	P209	4	9/8, 9/28, 9/29	9/8a, 9/28a, 9/29a
	P277	5	210/14, 9/10c	9/10c
Home Oil (UK) Ltd.	P211	4	210/19	210/19
London and Scottish Marine UK Co.	P333	7	3/27	3/27
Mobil Producing North Sea Ltd.	P139	4	9/13	9/13a
	P170	4	205/23, 205/29, 206/9, 206/10	205/23a, 206/9, 206/10a
	P312	6	209/3, 209/13	209/3, 209/13
	P337	7	9/13b	9/13b
Monsanto Ltd.	P174	4	204/30	204/30a
Occidental Petroleum (Caledonia) Ltd.	P250	4	9/6, 210/29	210/29a
	P332	7	3/21	3/21
Phillips Petroleum	P175	4	205/25, 206/21	205/25a, 206/21a
	P226	4	3/22, 9/16, 210/15, 211/1	3/22a, 210/15a, 211/1a
Placid Oil Co. (UK)	P212	4	211/3, 211/8	211/8a
Quintana Petroleum	P228	4	9/17	9/17a
Ranger Oil (UK) Limited	P229	4	3/30, 4/21, 4/26	3/30a, 4/21, 4/26

Operator	Licences	Round	Block Licensed Originally	Blocks Currently Licensed
Shell (UK) Ltd.	P117	3	9/23, 9/24, 211/29, 211/30	Part 211/29 (Licence rearranged)
	P177	4	207/2, 206/4, 205/15, 205/17, 205/18, 205/27, 205/28	205/18a, 205/27a, 207/2a, 205/28a, 206/4, 205/15a
	P231	4	210/4, 210/9	210/4a, 210/9a
	P232	4	3/12, 210/20, 210/25, 211/13, 211/14, 211/16, 211/21, 211/23, 211/26	Parts of 211/21a, 211/23a, 211/26a
	P257§	3		Part 211/29, 211/30, 9/23a, 9/24a
	P258	4		Part 211/21a
	P296	4		3/12a, 210/25a, 211/13a, 211/14, 211/16a, part 211/21a, part 211/26a, parts 211/23a
	P321	6		209/7, 206/2
Sovereign Oil & Gas Limited	P182	4	205/30	205/30a
	P330	7	3/11b	3/11b
	P331	7	3/16	3/16
Texaco North Sea (UK) Limited	P119	3	3/4	3/4a
	P183	4	207/1	207/1a
Total Oil Marine Limited	P090	2	3/9, 3/10, 3/14, 3/15 3/24, 3/25, 9/9, 9/10, 9/14, 9/15	3/9a, 3/10a, 3/14a, 3/15, 3/24a, 3/25a, 9/9a, 9/10a, 9/10b, 9/14a, 9/15a
	P118	3	3/20, 10/1	3/20a, 10/1
	P181	4	206/5	206/5a
	P239	4	3/1, 3/6, 3/19, 211/17, 3/18	3/1a, 3/6a, 3/19a, 211/17a, 3/18a
	P284	5	3/14b	3/14b
Tricentrol North Sea Limited	P316	6	208/23	208/23

Operator	Licences	Round	Block Licensed Originally	Blocks Currently Licensed
Unocal Exploration	P242	4	2/5, 9/12	2/5
	P254	4		9/12a
	P335	7	9/11	9/11
Zapata International Corporation	P208	4	3/13	3/13a

Addendum: **Licence Block Awards Offshore Shetland made on 12th March, 1981**

Total No. of Licences awarded: 12

Block No.	Operator
1/4	Tricentrol North Sea
206/1	British National Oil Corporation
208/13	Shell UK Ltd.
208/17	Amoco UK Petroleum Ltd.
209/4	British Sun Oil Co. Ltd.
209/12	Marathon Oil UK Ltd.
214/27	Gulf Oil Corporation
214/28	Esso Petroleum Co. Ltd.
219/20	Conoco (UK) Ltd.
219/27	Sovereign Oil and Gas Ltd.
219/28	Sovereign Oil and Gas Ltd.
220/16 220/21 220/22	Conoco (UK) Ltd.

THE VARLEY ASSURANCES

The "Varley Assurances" were oilfield depletion guidelines as laid down by Eric Varley the former Secretary of State for Energy, in 1974. These guidelines gave operators undertakings that:

1. no delays would be imposed on finds made up to the end of 1975;

2. no cuts would be imposed from these fields until 1982 or four years from the start of production;

3. no cuts would be imposed on post-1975 finds until 150 percent of the investment in the field had been recovered;

4. the Government would generally limit any cuts to 20 percent of planned daily output.

The object of this policy would be to reduce the level of potential net oil exports from Britain in the 1980's and to extend net oil self sufficiency as long as possible into the 1990's.

1 EXPLORATION WELLS DRILLED IN SHETLAND WATERS

	1971	1972	1973	1974	1975	1976	1977	1978	1979
East of Shetland*1	4	8	16	26	23	25	24	11	4
West of Shetland*2	—	1	—	8	3	1	11	1	3
Total Shetland (East & West)	4	9	16	34	26	26	35	12	7
Total UK Sector (including Shetland)	24	33	42	67	79	58	67	37	16
Percentage of UK Sector wells drilled in Shetland waters	16.67	27.27	38.10	50.75	32.91	44.83	52.23	32.43	43.75

NOTES: *1 East of Shetland: North of Latitude 59°N.
*2 West of Shetland: Area west of Shetland as far south as Block 156.

2 APPRAISAL WELLS*1 DRILLED IN SHETLAND WATERS*2

	1972	1973	1974	1975	1976	1977	1978	1979
East of Shetland	1	7	24	24	13	17	8	8
West of Shetland	—	—	—	—	—	—	3	—
Total of Shetland	1	7	24	24	13	17	11	8
Total UK Sector	8	19	33	37	28	38	25	16
Percentage of UK Sector wells drilled in Shetland waters	12.5	36.84	72.73	64.86	46.43	44.73	44	50

NOTES: *1 An appraisal well is one appraising an area where oil or gas have previously been discovered during exploration drilling.

*2 Waters north of latitude 59°N, in the North Sea are considered 'Shetland Waters' here. No appraisal wells were drilled west of Shetland in the period 1972-197 .

*3 No appraisal wells were drilled in Shetland waters before 1972.

3 DEVELOPMENT WELLS*1 DRILLED IN SHETLAND WATERS

	1975	1976	1977	1978	1979
East of Shetland	1	10	29	54	66
Total UK Sector	21	54	96	96	107
Percentage of UK Total in Shetland waters	4.76	18.52	30.20	56.25	61.68

NOTE: *1 Development wells are those drilled to produce oil and gas for commercial purposes. No development wells were drilled in Shetland waters before 1975.

4 TOTAL NUMBER OF WELLS*1 DRILLED IN SHETLAND WATERS

	1971	1972	1973	1974	1975	1976	1977	1978	1979
Wells drilled in Shetland waters	4	10	23	58	51	49	57	77	81
Total wells drilled in UK waters	62	77	82	120	137	140	201	158	139
Percentage of UK total wells drilled in Shetland waters	6.45	12.99	28.05	48.33	37.23	35.00	28.35	48.7	58.25

NOTE: *1 Total wells = Exploration wells plus Appraisal wells plus Development wells.

5 MOBILE DRILLING RIG ACTIVITY OFFSHORE SHETLAND

Rig time — Measured in rig years*1

	1971	1972	1973	1974	1975	1976	1977	1978	1979
East of Shetland	0.8	2.7	6.9	12.4	13.6	9.9	10.1	5.9	4.5
West of Shetland	—	0.1	—	1.5	0.3	0.4	1.8	1.1	0.4
Total Shetland	0.8	2.8	6.9	13.9	13.9	10.3	11.9	7	4.9
Total UK Sector	5.2	8.8	13.3	24.5	27.7	21.2	29.9	18.1	16.1
Percentage Shetland activity of UK Sector total	15.38	31.82	51.88	56.73	50.18	48.58	49.79	38.6	30.43

NOTE: *1 Rig Year: Time equivalent to one rig drilling non-stop for one year. (On average it takes a rig 60 days to drill a well).

6 FIXED PLATFORM ACTIVITY OFFSHORE SHETLAND

Rig Time — measured in rig years

	1975	1976	1977	1978	1979
East of Shetland*1	0.2	3.2	6.0	11	14.5
Total UK Sector	2.6	9.3	14.9	18.6	21.5
Percentage Shetland activity of UK Sector Total*2	7.69	34.41	40.26	59.1	67.44

NOTES: *1 No fixed platforms have been placed yet to the West of Shetland.

*2 The percentage of fixed platform activity offshore Shetland compared to the UK total is expected to continue to increase.

WELLS DRILLED WEST OF SHETLAND

Well	Operator	Date of[1] Spudding	Date Completed	Comments
206/12-1	Esso	11.7.72	8.8.72	P&A[2] oil shows
205/21-1	Shell UK	25.1.74	28.5.74	Indications of hydrocarbons: Test efforts unsuccessful. Surrendered 17.06.77
202/2-1	Esso	13.6.74	29.6.74	P&A
202/8-1	Shell UK	23.6.74	16.7.74	P&A Surrendered 17.6.77
205/20-1	Elf Oil Explo.	17.7.74	28.8.74	P&A
207/2-1	Shell UK	21.7.74	6.9.74	P&A; dry
205/22-1	BP Pet. Dev.	2.8.74	26.12.74	Abandoned; surrendered 1978
202/3-1	BP Pet. Dev.	13.10.74	18.1.75	P&A; dry
205/30-1	Siebens Oil & Gas	31.8.74	8.10.74	P&A
205/23-1	Mobil	12.3.75	19.4.75	P&A; surrendered 1978
205/26-1	Arco	18.6.75	25.7.75	Dry and abandoned; relinquished 1978
204/30-1	Monsanto	27.7.75	22.8.75	P&A
206/5-1	Total/Shell	22.4.76	3.9.76	P&A
202/3-2	BP Pet. Dev.	13.4.77	8.5.77	No information
206/11-1	Elf	7.5.77	9.10.77	P&A; non-commercial gas shows
206/8-1	BP Pet. Dev.	10.5.77	14.5.77	Junked at 453 ft[3]
206/8-1a	BP Pet. Dev.	15.5.77	29.7.77	Flowed 2920 b/d[4] (24°- 25°) API oil through 1 inch choke from two intervals at 452.7 ft
206/12-2	Esso	5.8.77	5.10.77	Wildcat P&A 23°- 25° API oil five cased hole prod. tests. max 630 b/d through 24/64 inch choke. Av. well pressure 273 lb. Very viscous oil; negligible GOR
207/1-1	Texaco	18.8.77	28.8.77	P&A
207/1-2	Texaco	4.9.77	27.9.77	P&A
207/1-3	Texaco	29.9.77	29.10.77	Minimal testing at 517 ft; gas shows
206/9-1	Mobil	5.10.77	19.12.77	Wildcat. P&A dry

WELLS DRILLED WEST OF SHETLAND — *Continued*

Well	Operator	Date of[1] Spudding	Date Completed	Comments
205/25-1	Phillips	12.10.77	22.12.77	P&A; tight hole[5]
206/7-1	Elf	10.11.77	10.2.78	Flowed 1700[6] b/d at 23° API oil[7] with low GOR and 3.9m (10^6) cfd gas from ⅜ inch choke; three intervals at 494.4 ft
206/13-1	Esso	3.12.77	12.1.78	P&A; tight hole; relinquished 1978
206/9-2	Mobil	6.1.78	2.3.78	P&A; tight hole
206/8-2	BP Pet. Dev.	27.4.78	19.8.78	Tested gas. Produced oil from 22° to 25° API at non-commercial rates
206/8-3A	BP Pet. Dev.	12.9.78	4.12.78	206/8-3; 19.8.78-11.9.78; junked. Oil shows 530 b/d from lower interval. Two tests. Upper interval unproductive
206/8-4	BP Pet. Dev.	20.12.78	11.2.79	Exploration well 24° API oil found but P&A with no testing
208/27-1	BNOC	16.5.79	18.6.79	Wildcat; tight hole
208/15-1A	BP/BNOC	17.8.79	14.11.79	P&A; tight hole
209/9-1	BNOC	4.10.79	2.1.80	P&A; tight hole
209/6-1	Gulf	12.3.80	7.7.80	P&A; tight hole
206/2-1	Shell	8.4.80	11.10.80	Drillship Petrel drilled in 2007 ft water depth; tight hole
206/10a-1	Mobil	30.4.80	4.7.80	Wildcat, P&A, tight hole
206/8-5	BP Pet. Dev.	18.5.80	21.7.80	Appraisal well, potential oil producer. Hydraulic fracture treatment increased flow rate from 800 b/d to 2000 b/d for same well head flow pressure
206/8-6	BP Pet. Dev.	23.7.80	1.9.80	P&A
209/3-1	Mobil	19.8.80	23.10.80	Wildcat; tight hole

NOTES:
- [1] Spudding = Commencement of drilling
- [2] P&A = Plugged and Abandoned
- [3] Junked = Abandoned because of technical difficulties
- [4] b/d = Barrels of oil per day
- [5] Tight hole = No information being given out by oil companies
- [6] GOR = Gas/oil ratio
- [7] Scfd = Standard cubic feet per day

SUMMARY: Total wells drilled to end 1980 = 38

NAMED FIELDS WITH UK GOVERNMENT APPROVED DEVELOPMENT PLANS OFFSHORE SHETLAND

Chart I

Name of Field	Licence Block No.	Discovery Date	Water Depth (ft)	Operator	Platforms No. and Type	Installation Date	Production Startup	Peak Production (Barrels oil per day)	Estimated Recoverable Reserves (Million Barrels)	Method of Transportation	Comments
South Beryl 'A'	Location 9/13a approx. 88 miles to SE of Sumburgh	Sept. 1972	385	Mobil	1 Concrete	1975	June 1976	115,000 in 1980	400; gas/oil ratio (GOR) 815-1300	Off-shore tanker loading via exposed location single buoy mooring (ELSBM)	In 1980 'A' achieved 100 million barrels production since output began in 1976. There are two other known adjacent reservoirs in Beryl area:— 1. *North Beryl or Beryl 'B'*. There is understood to be no pressure connection between Beryl 'A' and Beryl 'B' structures, but they are contiguous and are treated as one field for administration and tax purposes (see Beryl 'B' below). 2. *West Beryl* 9/13-4, discovered June 1976, with possibly 60 to 100 million barrels recoverable reserves of oil, which may be developed using subsea technology.
North Beryl (Beryl 'B')	9/13a	May 1975	390	Mobil	1 Steel	1983	1984	85,000 in 1985 for five years	300	Oil via pipeline to Beryl A Platform (to be laid in 1983) then tanker loading via ELSBM	Well 9/13-7 identified Beryl B in 1975 and test production began early 1979 from well tied into Beryl A Platform via subsea completion. Development plans for £600 million project including steel platform five miles north west of Beryl A, were submitted to Dept. of Energy late 1979. Seabed template installed summer 1980 for pre-drilling up to six wells.
Brent (A, B, C, D)	211/29, 3/4 approx. 114 miles to NE of Sumburgh	July 1971	460 ft	Shell/ Esso	4:— A Steel B Condeep C Concrete D Condeep	May 1976 Aug. 1975 June 1978 June 1976	June 1978 Nov. 1976 1981 Nov. 1977	450,000 in 1984	1685 Stabilised Crude Oil; 530 NGL + 3 trillion cu. ft. gas	1. *For Crude Oil* — Initially spar buoy tanker loading; also from Nov. 1979 via £90 million, 96 mile long, 36 inch diameter joint user pipeline to Sullom Voe from Cormorant commissioned for carrying oil from Brent Field. All Brent oil for pipeline is pumped to Brent C then via Cormorant A Platform and a further 96 miles to Sullom Voe. Pipeline capacity 1 million barrels per day. 2. *For Gas and NGL* some transported in solution to Sullom Voe from 1981 when gas processing facilities commissioned; 85% or more of Brent gas to go via £300 million Far North Liquids and Associated Gas System (FLAGS) 281 mile long 36 inch diameter pipeline to St. Fergus, laid by barge SEMAC from March 1977 to June 1978, this pipeline of 1.1 billion cu. ft./day (at 2000 P.S.I.) capacity, was completed after 324 days work.	Capital investment in Brent Field complex is in excess of £3000 million. Brent Field has highest GOR of fields currently under development in East Shetland Basin and five of earlier wells on platforms A & B were shut in on account of excessively high GOR. In 1980 hook up of water injection and gas compression equipment continued, perhaps spurred on since UK government tightened restrictions on gas flaring in Nov. 1979. Flaring restrictions contributed to Brent A, B, D, production cut from 200,000 to 120,000 BOPD — Brent C due on stream mid-1981.

NAMED FIELDS WITH UK GOVERNMENT APPROVED DEVELOPMENT PLANS OFFSHORE SHETLAND — Continued

Chart 1

Name of Field	Licence Block No.	Discovery Date	Water Depth (ft)	Operator	Platforms No. and Type	Installation Date	Production Startup	Peak Production (Barrels oil per day)	Estimated Recoverable Reserves (Million Barrels)	Method of Transportation	Comments
South Cormorant ('A')	211/26a approx. 102 miles to NE of Sumburgh	Sept. 1972	429	Shell/Esso	1 Concrete	May 1978	Dec. 1979	60,000 in 1982	110; GOR 500-650	Oil via Brent system pipeline to Sullom Voe directly from Cormorant 'A' Platform Gas line to Brent and FLAGS	Cormorant 'A's first oil production via Brent system to Sullom Voe was in Dec. 1979. Cormorant Field encompasses structures to be drained by three separate production systems: viz. 1 South Cormorant (A) 2 North Cormorant (see below) 3 Central Cormorant — this structure is expected to be linked to a subsea manifold. Cormorant A is crucial to Brent pipeline system containing pumps, pig receiving station, telecommunications and other facilities reflecting its key position.
North Cormorant	211/21a	July 1974	525	Shell/Esso	1 Steel	1981	1982	180,000	400	Oil via 20 inch diameter pipeline to Cormorant A platform thence to Sullom Voe via 36 inch Brent pipeline gas via 10 inch feeder line to FLAGS.	Development approval for £700 million project given by UK government in April 1979. Platform to come from UIE, Cherbourg, France.
Dunlin	211/23a approx. 119 miles NE of Sumburgh	July 1973	495	Shell/Esso	1 Steel/Concrete Hybrid	June 1977	Aug. 1978	150,000 in 1982	450-600 (being reassessed); GOR 250	24 inch diameter pipeline to Cormorant 'A' Platform thence to Sullom Voe.	Oil from Dunlin via Brent system pipeline was first oil ashore at Sullom Voe on November 25th 1978.
Hutton	211/28, + 211/27 approx. 105 miles to NE of Sumburgh	Nov. 1973	485	Conoco	1 Tension Leg Platform (TLP)	1983	1984	115,000	260; Low GOR	Option to join Brent System (oil and gas)	Development approval for £600 million project given by Dept. of Energy 13th Aug. 1980. First field to be developed by TLP which new technology may be applied to other deeper water sites in future.
Hutton Northwest	211/27	April 1975	470	Amoco	1 Steel	Aug. 1981	Sept. 1982	100,000 + 35 million cu. ft./day of gas in 1983	280; GOR 300	Oil via 20 inch diameter pipeline to Cormorant 'A' Platform, thence via 36 inch Brent pipeline to Sullom Voe. Gas and associated gas liquids via 10 inch diameter spur line to 16 inch Western FLAGS line.	£500 million development plans approved by UK government July 1979. By October 1979 seabed template installed and pre-production drilling commenced. Platform being built by McDermott, Ardersier.
Murchison	211/19 + Norwegian Sector 33/9 approx. 127 miles NE of Sumburgh	Sept. 1975	512	Conoco	1 Steel	Aug. 1979	Sept. 1980	150,000 in 1982	350/380	Oil via 16 inch diameter 10 mile long pipeline to Dunlin Field thence into Brent system to Sullom Voe. Gas reinjected initially.	£500 million project. Approximately 83 percent of this field lies in UK North Sea; the rest is in the Norwegian Sector.

NAMED FIELDS WITH UK GOVERNMENT APPROVED DEVELOPMENT PLANS OFFSHORE
SHETLAND — Continued
Chart I

Name of Field	Licence Block No.	Discovery Date	Water Depth (ft)	Operator	Platforms No. and Type	Installation Date	Production Startup	Peak Production (Barrels oil per day)	Estimated Recoverable Reserves (Million Barrels)	Method of Transportation	Comments
Thistle	211/18a approx. 124 miles to NE of Sumburgh	July 1973	525	BNOC	1 Steel	Aug. 1976	Feb. 1978	210,000 in 1980	450 + 106 in 211/18-6 area + 2.1 trillion cu. ft. gas (gas to be reinjected)	Offshore loading by Single Anchor Leg Mooring (SALM) and also oil via 16 inch diameter pipeline to Dunlin thence Cormorant 'A' Brent system pipeline to Sullom Voe.	Possibility of BNOC developing other oil accumulations in this vicinity in areas 1, 6, 9, 12 and 13 in 211/18a.
Heather	2/5 approx. 94 miles NE of Sumburgh	Dec. 1973	470	Union Oil	1 Steel	May 1977	Oct. 1978	50,000 in 1982	150; GOR 630	Oil via 16 inch diameter pipeline to Ninian Platform thence via 36 inch diameter Ninian system pipeline to Sullom Voe. Possibility gas may go into western leg of FLAGS.	First oil ashore through Ninian pipeline system to Sullom Voe was from Heather on 3rd December 1978. West Heather — structure discovered on 2/5 in 1974 with a possible 75 million bbls estimated recoverable reserves, may be developed in future. In September 1979 Union Oil found more oil 7½ miles SW of Heather on 2/5. This is also another possibility being evaluated in this area.
Magnus	211/12 211/7 approx. 130 miles to NE of Sumburgh	June 1974	620	BP	1 Steel	1982	1983	120,000 + 50 million cu. ft./day gas + 9000 b/d gas liquids	484	Oil via 24 inch diameter 92 km long pipeline to Ninian Central Platform thence via 36 inch diameter Ninian system pipeline to Sullom Voe. Gas probably via Northern leg pipeline feeding into FLAGS or new gas gathering pipeline system.	Most northerly UK field yet developed extending to 61° 40'' North. £1315 million project. Most of field in 211/12 with northern end in 211/7. Drilling expected to be completed by 1986. Pipeline construction to start spring 1981, will include 6 inch diameter flowlines to seven subsea satellite wells in addition to 92 km line to Ninian Central Platform.
Ninian	3/3 3/8a approx. 101 miles to NE of Sumburgh; and 73 miles from Unst	Jan. 1974	445.2 to 464.2	Chevron	3:— 1 Steel (3/8) 1 Concrete (3/3) 1 Steel (3/3)	May 1977 May 1978 June 1978	Dec. 1978 June 1979 Sept. 1980	325,000 in 1981 for two years	1000 + 65m bbls NGL + .24 billion cu. ft. gas. GOR 380	Oil — 105 mile long, 36 inch Ninian system pipeline from Ninian Central Platform to Sullom Voe. Gas via western leg of FLAGS?	Ninian is a £1400 million, three platform development, platforms accounting for £880 million of this total investment. Field is one of largest in North Sea, being approx. four miles wide, 14 miles long. Chevron has had further discoveries on 3/7 and 3/13, south and west of Ninian. To west is Alwyn Field (operated by Total) — it is possible any or all of these finds could be linked into Ninian system to maximise use of platform and pipeline facilities.

NAMED FIELDS WITH NO DEVELOPMENT PLANS YET ANNOUNCED

Chart II

Name of Field	Operator/Owner	Licence Block No.	Discovery Date	Water Depth (ft)	Remarks
ALWYN	Total Oil Marine Ltd.	3/14a	November 1973	430	Four appraisal and two wildcat wells drilled to date — the last being in Sept 1977. First wildcat drilled in 1973 resulted in a flow rate of 3000 BOPD of 42° API oil with associated gas through a half inch choke and indicated excellent potential. Thereafter, appraisal drilling indicated a varied and complicated structure. Estimate of reserves of 182 million bbls oil (and 15-30 billion cu. metres gas). No development plans announced as Total have given North Alwyn priority but thought development would be by subsea system to North Alwyn.
ALWYN (North)	Total Oil Marine Ltd.	3/9a (Approx. 100 nautical miles east of Lerwick)	November 1975	420	In June 1980 Total announced intention to develop this field. Estimated reserves are up to 200 million bbls crude oil and 30 billion cu. metres gas. Development expected to cost £1.5 billion and to include two fixed steel platforms. Reserves are in three separate reservoirs — one of oil and gas, two of gas only. Production has been estimated at 40.000 BOPD for eight years, production start up probably 1986. Oil will probably go to Sullom Voe through either the Brent or Ninian Pipeline systems, gas may go either to Frigg Field 50 miles south of Alwyn and then to St Fergus or else to new gas gathering pipeline system (see appendix). First platform to be installed 1983.
BRUCE	Hamilton Brothers	9/8 (Approx. 80 nautical miles south east of Lerwick)	February 1975	380	Gas and condensate discovery with possible reserves of 1000 billion cu. ft. gas and up to 250 million bbls of associated NGL. Investment needed to develop provisionally estimated at £500 million. Transportation could be via proposed North Sea Gas gathering pipeline system. No development will proceed without transportation facilities being available. Further delineation drilling programme in progress in this area by both BP and Hamilton Brothers.

CLAIR	BP Petroleum Development Ltd.	206/8 (Approx. 34 miles north west of Sullom Voe, Shetland)	July 1977	Five wells completed to date (two wells, 206/8-1 and 206/8-3 were junked). 2068-1a had aggregate flow of 2920 BOPD 24-25° API through one inch choke, to total depth 7631.6 ft; 206/8-2 tested gas from small accumulation at top of reservoir and four deeper intervals produced oil from 22 to 25° API at non commercial rates. 206/8-4 encountered some 24° API oil of similar quality to other wells in the area, not tested; 206/8-5 was subjected to fracture testing which increased flow from 800 to 2000 BOPD. This field is not expected to have development plans submitted in near future as viscosity of oil will require new technology to enhance recovery. Structure is thought to straddle 206/7, 206/8, 206/9, 206/12 and 206/13. Some estimates have suggested a vast oilfield up to 4 billion barrels oil in place.
EIDER	Shell UK Exploration & Production Ltd.	211/16 (Approx. 94 nautical miles north east of Lerwick)	June 1974	Five wells drilled to date: two dry, three oil wells. Further appraisal expected. Shell will produce nearby Tern Field first. Recoverable reserves estimated at 120 million bbls; production probably 1987.
LYELL	Conoco	3/2 (Approx. 85 nautical miles north east of Shetland)	June 1975	Four oil wells drilled to date. 3/2-1a drilled in 1975 tested at rates between 1000 and 4800 BOPD from three separate zones of middle Jurassic sand which is the major producing formation for fields in the area; 3/2-3, delineation well 2½ miles north of 3/2-1a tested oil in lower Brent sands at rates of up to 414 BOPD through ¼ inch choke; upper horizons indicated productive, cored, not tested. Estimate could be up to 150 million bbls recoverable reserves but Conoco still evaluating the structure.
TERN	Shell/Esso	210/25 (Approx. 90 nautical miles north east of Lerwick)	April 1975	Recoverable reserves estimated at 140 million bbls. Feasibility study in hand, if feasible, production scheduled 1985, probably via conventional fixed steel platform, with oil produced into Brent, system pipeline to Sullom Voe via North Cormorant Platform 15 kilometres to South East and associated Gas produced into Western Leg gathering system (WELGAS) to go via FLAGS to St Fergus. Four wells drilled to date, last well drilled in 1977.
WENDY	Phillips	210/15 (Approx. 105 miles north east of Lerwick)	October 1977	No immediate plans — maybe developed late 1980's; two wells to date. 210/15-2, 37.6° API oil discovered, which flowed at maximum rate of 4760 BOPD through one inch choke.

UNNAMED DISCOVERIES OF INTEREST OFFSHORE SHETLAND

Chart III

Licence Block No.	Location	Discovery Date	Water Depth (Feet)	Operator	Comments
2/5-10	Approx 75 miles NE of Lerwick	September 1979	490	Union Oil	This is a separate structure from the Heather Field which it is close to. The exploration well flowed oil at 6000 BOPD. Further appraisal expected.
2/10a	66 nautical miles NE of Lerwick; due south of Heather Field	April 1975	517	Chevron	Four wells to date. 41° API encountered on 2/10-1a (2/10-1 abandoned due to drilling difficulties). Possibly there could be a structure containing 100 million bbls of recoverable reserves or less. Second well dry; third well had oil and gas shows. Fourth well flowed 21-24° API oil at 2325 BOPD on 1¾ inch choke and 696 BOPD on 1¼ inch choke in 2 drill stem tests. This is thought to be a separate structure with up to 150 million bbls recoverable reserves.
3/4	110 miles NE of Lerwick; due south of Brent Field	October 1973	447	Texaco	Seven wells to date; last drilled 1975: First well, 3/4-1 flowed 36° API oil, rates ranging from 2185 to 6139 BOPD through various chokes. Third well 3/4-2a flowed 35.7° API oil at 7022 BOPD through ¾ inch choke. In a separate fault segment fourth well flowed 41.5° API oil at 5785 to 7485 BOPD with various chokes. Fifth well, 3/4-5, flowed 42.5° API oil at 4578 BOPD, GOR 5760 cu. ft./bbl. 3/4-6 penetrated Statfjord formation, flowed on 5/8 inch choke at combined rates over two intervals of 46° API oil at 5534 BOPD. av. GOR 10390 cu. ft./bbl. Third test flowed 821 BOPD oil in Brent Sands, GOR 702 cu. ft/bbl. Seventh well 3/4-7, plugged and abandoned.
3/7 and 3/8-5A	75 miles NE of Lerwick; adjacent to Ninian Field	June 1976 and March 1980	479 to 450	Chevron and BP	3/7: Three wells drilled to date. First well, 3/7-1, discovered 37.9° API oil which flowed at 3100 BOPD. Second well 3/7-2, had 34.9° API oil flow at max. rate of 3300 BOPD. Third well was plugged and abandoned. 3/8a-5A: This is thought to be part of the same structure as the 3/7 find. This well flowed at 3688, 100, 4874 and 6520 BOPD in separate tests.
3/8a-6	Approx 75 miles NE of Lerwick	June 1980	480	BP	This is the third separate structure found in the Ninian area. Oil flowed at 7200, 4900 and 3700 BOPD at four separate intervals.
3/9a	70 nautical miles NE of Lerwick; adjacent to Ninian Field	October 1975	411	Total	Four wells to date. First well, 3/9a-1, flowed 2717 BOPD through ½ inch choke; second well was oil well; third well plugged and abandoned after gas condensate found. 3/9a-4, adjoining Alwyn Field, flowed 38° API oil at 6173 and 6139 BOPD.

Block	Location	Date	Water depth (ft)	Operator	Comments
3/28	77 nautical miles E of Lerwick	May 1976	306	Chevron	Two wells to date. First well 18.5° API oil, difficult to test; second well 3/28-2 pumped 3400 BOPD of 11.7° API oil. Not commercially viable yet, with current technology.
206/7	40 nautical miles WNW of Sullom Voe	February 1978	495.4	Elf	Tight hole. Flowed 23° API oil at 1700 BOPD aggregate rate and 3.9 million cu. ft./day gas on 3/8 inch choke from three test intervals. Elf stated in February 1981 that pilot oil production being considered on this structure, which is thought to be linked to 206/8 (Clair Field), 206/9, 206/12 and 206/13.
206/9	32 nautical miles WNW of Sullom Voe		442	Mobil	First well, 206/9-1 drilled 1977, plugged and abandoned as dry hole. Second well 206/9-2 drilled 1978, plugged and abandoned. Tight hole status.
206/12	37 nautical miles W of W mouth of Sullom Voe	October 1977	570	Esso	Two wells to date. First well drilled 1972 plugged and abandoned. Second well 206/12-2, tight hole, flowed 23-25° API oil in five cased hole production tests at a max. rate of 630 BOPD through 24/64 inch choke; av. well head pressure 273 lb.
206/13	33 nautical miles WNW of Sullom Voe		416	Esso	One well only to date, drilled December 1977/Jan 1978. Plugged and abandoned. Tight hole status.
211/13	108 nautical miles NE of Lerwick; due east of Magnus Field	July 1971	585	Shell/Esso	Seven wells to date — oil shows in three of them. Substantial gas and condensate in first well. Last well drilled (in 1978) was dry.
211/18-12		July 1976	554	BNOC	Wildcat — separate geological feature from Thistle Field. Flowed 39.7° API oil at 1529 to 2932 BOPD through restricted chokes. Zone 12, sometimes called Caber, North Thistle or North East Thistle. 140 million barrels of recoverable reserves.
211/18-13		September 1976	537	BNOC	Stepout from 211/18-12. 42° API oil flowed at approx. 9100 BOPD.
211/19-6	106 nautical miles NE of Lerwick	February 1977	512	Conoco/BNOC/Gulf	211/19-6 wildcat flowed at 5500 BOPD; further studies needed to establish possible commercial viability. Seventh well dry.

Appendix to Graph B

FIGURES OF PRODUCTION FROM UKCS OILFIELDS ON WHICH GRAPH BASED

Year	12 Fields North of 59°N ('000 B/D)	Total UKCS	% of Total UKCS[1] North of 59°N
1976	11	238	4.62
1977	87	764	11.38
1978	203	1081.5	18.77
1979	646	1574	41.04
1980	758	1637.5	46.29
1981	1007	1981.5	50.82
1982	1304	2321	56.18
1983	1490	2480	60.08
1984	1637	2687.5	60.91
1985	1739	2752	63.19
1986	1664	2667.5	62.38
1987	1454	2476	58.72
1988	1395	2255.5	61.84
1989	1208	1975.5	61.14
1990	1042	1730	60.23
1991	863	1450.5	59.49
1992	761	1207	63.04
1993	663	1002.5	66.13
1994	561	828	67.75
1995	458	642.5	71.28
1996	378	525	72
1997	298	410	72.68
1998	239	301.5	79.27
1999	195	230.5	84.59
2000	150	150.5	99.66
2001	97	97.5	99.48

*FOOTNOTE: *1 The percentage of production from fields North of 59°N increases with time due to the fact the northern fields have been later in developing*

GAS GATHERING

Gas Gathering schemes for the Northern North Sea have been considered since at least 1974, if not earlier. There have been four reports of interest produced on possible schemes, viz:

(i) in 1975, by Consultant Chemical Engineers, Buchanan & Clacher, and updated in 1977;

(ii) in 1976, by Williams Merz — commissioned by the UK Government;

(iii) in 1978, by Gas Gathering Pipelines (North Sea) Limited for the UK Government;

(iv) in 1980 by the British Gas Corporation and Mobil North Sea Limited, for the UK Government.

(i) **Buchanan & Clacher** suggested two alternative pipeline networks:
  (a) A common carrier gas pipeline, running from North to South, close to the median line with Norway, from Statfjord via Brent and Alwyn Fields, to turn south-west to parallel the Frigg pipeline route to St. Fergus with intermediate stops being provided en route to pick up additional gas and boost pressure;
  (b) Additional facilities to transport gas from fields with smaller amounts of associated gas to the Scottish Mainland. Gas lines would run from the Statfjord, Alwyn and Brent Fields, thence to the Shetland Islands, the Orkney Islands and the Scottish Mainland.

(ii) **Williams-Merz** suggested a four pipeline system approximately 800 miles long. Amongst other things the Brent-Magnus complex would be linked either to the Brent pipeline or to a central system pipeline passing through the Alwyn, Frigg and near the Claymore Fields to St. Fergus.

By 1977 the Scottish Development Department isolated four possible landfall sites for gas pipelines in Scotland: Shetland, Orkney, the Inner Moray Firth,* and the Rattray Head/St. Fergus area.** Orkney and Shetland were later discarded as possible landfalls because of their remoteness.

Meanwhile in December 1976, the UK Government decided to establish a study company to:
(1) prove the viability of a gas gathering scheme;

(2) select the best route(s) for a pipeline scheme;

(3) consider the organisational and financial needs of such a project

This study company, Gas Gathering Pipelines (North Sea) Limited (GGP) comprised the British National Oil Corporation (BNOC), the British Gas Corporation (BGC), ELF/Total, Rio Tinto Zinc, and Imperial Chemical Industries, BNOC and BGC having one-third shares in the Company. GGP submitted its report in April 1978. Its main conclusion was that, at that stage, it appeared that insufficient gas would become available from the UK sector of the North Sea to justify a new trunk line to the shore. The Report recognised the situation might change if large additional sources of gas were discovered and developed in the UK sector or if account was taken of gas from the Norwegian sector.

---

*Inner Moray Firth is where Cromarty Petroleum have land with approved development plans.

**St. Fergus is where the Brent FLAG's and the Frigg gas pipelines have their terminals.

FIELDS WITH GAS FOR COLLECTION OFFSHORE SHETLAND
(commercial and larger discovered fields)

Licence Block	Name of Field (if given)	Operator
211/12	Magnus	BP
211/13		Shell
211/18	Thistle	BNOC
211/19	Murchison	Conoco
211/23	Dunlin	Shell
211/26	Cormorant	Shell
211/28	Hutton	Conoco
2/5	Heather	Unionoil
2/10		Siebens
211/24	Statfjord	Conoco
3/3	Ninian	Chevron
3/4		Texaco
3/8		BP
3/14	Alwyn	Total
3/25		Total
9/8	Bruce	Hamilton
9/13-7		Mobil
9/13	Beryl	Mobil

Source: Williams-Merz study, 1976

By Spring 1979, the UK Government decided the situation had changed sufficiently to justify a further investigation into the feasibility of a gas gathering system. The increasing price of oil had revived development hope for marginal fields* and the growing shortage of energy emphasised the importance of conserving gas that might be flared.

In July 1979, the UK Government therefore invited the British Gas Corporation (BGC) and Mobil North Sea Limited (Mobil) to undertake jointly a design and feasibility study into a gas gathering system to bring ashore gas and gas liquids from a number of fields in the Northern and Central sectors of the Northern North Sea. This report was submitted on 1st April 1980. Included in its recommendations were the following:

(a) A new offshore pipeline system should be built to collect gas from UK fields in the northern basin between latitudes 56° North and 62° North.

The offshore pipeline should be a 36 inch diameter trunk line originating near Statfjord and landing at St. Fergus via the Magnus, Murchison, Thistle, NE Thistle, UK Statfjord, Beryl A and B and South West Beryl, 9/10, 9/9, South Brae and the "T" Block Fields should be connected to this trunk line via lateral links or directly.

There was also to be a southern leg to the pipeline system.

(b) Provision should be made in the design to link up additional fields expected to come into production in the late 1980's and 1990's.

This pipeline system is scheduled to be ready for 1984/85 if current time scales are adhered to, and should stimulate further developments offshore Shetland.

THE RISE IN THE PRICE OF OIL* FROM THE UK NORTH SEA 1976-81

Date	Prices	Date	Prices
Jan-Mar 1976	$12.55	Jan-Mar 1979	$15.50
April-June 1976	12.70	April-May 1979	18.30
July-Sept 1976	13.20	June 1979	20.75
Oct-Dec 1976	14.10	July-Oct 1979	23.25
		Nov 1979	26.02
Jan-Mar 1977	$14.30		
April-June 1977	14.30	Jan 1980	$29.75
July-Sept 1977	14.30	Feb-Mar 1980	33.75
Oct-Dec 1977	13.98	April 1980	34.25
		May 1980	36.25
Jan-Mar 1978	$13.75		
April-June 1978	13.62	Jan 1981	$39.25
July-Sept 1978	13.72		
Oct-Dec 1978	14.00		

Note: *average UK North Sea Oil Price is used here.

(Source of information: British National Oil Corporation).

---

*A marginal field is one where the oil or gas reserves are not sufficiently large to make the economics of development a clear cut issue.

APPENDIX TO CHAPTER TWO

THE STRUCTURE OF LOCAL GOVERNMENT IN SHETLAND

The reform of local government was incorporated in the Local Government (Scotland) Act, 1973.

Local Government in Shetland was re-organised in 1974 when the Shetland Islands Council was elected. The Council began its executive functions in May 1975. Before that, the Zetland County Council, the Lerwick Town Council, and 14 District Councils were the bodies involved in local government in Shetland, under the terms of the Local Government (Scotland) Act 1947.

1. **The Zetland County Council**

    This consisted of 36 Councillors, 24 'Landward' or County members and the 12 members of the Lerwick Town Council, until 1972-75 when numbers were reduced to 33 by only nine Lerwick Town council members being on the Zetland County Council.

    The principal Committees of the Zetland County Council were:

    (a) **General Purposes Committee:** amongst other functions this Committee dealt with Planning matters.

    (b) **Health Committee:** which dealt with matters now under the auspices of Environmental Health and Control.

    (c) **Convener's Committee:** this met for special purposes and consisted of the Convener, Vice Convener, and all Committee Chairmen.

    (d) **Roads Committee.**

    (e) **Policy Committee:** this Committee was only formed in 1973.

    (f) **Landward Committee:** this Committee dealt with housing, water, drainage and other services, outwith the Burgh of Lerwick, where these matters were dealt with by the Town Council.

    (g) **Outer Isles Standing Committee:** for problems specific to the Outer Isles.

    (h) **Social Work Committee.**

    (i) **Development Committee.**

    (j) **Education Committee:** this included functions now designated as 'Leisure and Recreation'.

    (k) **Museum Management Committee.**

2. **The Shetland Islands Council**

    The Shetland Islands Council was elected in 1974, and took up executive functions in May 1975, with 24 members, which increased to 25 in 1978, after the Boundaries Commission had redrawn electoral areas or boundaries, in both the landward areas and in Lerwick. From May 1978, seven members represented Lerwick, instead of the 1975-1978 situation when there were only six.

    The Council is elected every four years — from the first election in 1974, there are elections every fourth year thereafter.

The principal Committees of the Shetland Islands Council are:

(a) **Social Work Committee.**

(b) **Ports and Harbours Committee:** for Sullom Voe and fishery harbours (piers were previously dealt with by the Landward Committee).

(c) **Environmental Health and Control Committee.**

(d) **Housing Committee:** (previously housing was a Landward Committee function).

(e) **Transport and General Services Committee:** (for roads, drainage, water,* transport, piers, etc.).

(g) **Development, Leisure and Recreation Committee.**

(h) **Resources Committee.**

(i) **Policy Committee.**

(j) **Chairman's Committee:** (this is the equivalent of the former Convener's Committee and meets as required for ad hoc matters).

In effect three new Committees — Ports and Harbours, Housing and the Development, Leisure and Recreation Committee — now exist. These Committees reflect changes brought about by the re-organisation of local government.

There is no Planning Committee, instead, since 1975, planning matters have been dealt with by meetings of the full Council, which takes place regularly once a month, when the Council is in session.

The staffing structures of the Council were also re-organised, and there is now a Management Team which meets regularly, consisting of the eleven Council Department Directors, headed by the Chief Executive.

Community Councils also came into being after the re-organisation of local government, and there are now 17 Community Councils in Shetland. These Community Councils differ from the former District Councils in that they have no local authority functions, all local authority functions now being vested in the all purpose authority** of the Shetland Islands Council. The Community Councils provide a useful service in their advisory capacity to the local authority, and act as a sounding board for any matters arising in their local communities. New areas encompassed by these local Councils are Lerwick, Sandwick and Scalloway — otherwise the areas covered by Community Councils are much as they were before re-organisation of local government under the District Councils, although their functions are different.

Prior to re-organisation this was the function of the North of Scotland Water Board.

Addendum: The Council's function as a Valuation Authority has been delegated to a joint Committee serving Orkney and Shetland, and the Assessor carries out his duties in both Islands' areas.

**Shetland, Orkney and the Western Isles are the only all purpose local authorities in Scotland. In the rest of the country local authority services are provided by Regional or District Councils.*

APPENDIX TO CHAPTER THREE

OIL RELATED IMPORTS FOR LERWICK HARBOUR

Year	Total (metric tonnes)	Coastwise*1 (metric tonnes)	Foreign*2 (metric tonnes)	Main Items
1971	342	N/A*3	N/A	only Barytes
1972	3626	N/A	N/A	only Barytes
1973	19807	5814	13993	Barytes, chemicals, cement, steel for Norscot Base construction
1974	47967	24932	23035	Barytes, chemicals, cement, steel and 3000 tonnes for bunkers at Norscot Base. (before 1974, no way of identifying oil related bunkers)
1975	255234	232796	22438	171000 tonnes of concrete coated pipe; 16000 tonnes bunkers, remainder mainly barytes, cement, chemicals
1976	161454	119236	42218	Concrete coated pipe over ⅓ of total tonnage imported; petrol products ¼ of total imports — all coastwise; crude minerals mostly from abroad; metal manufactures; casing, chemicals and cement mostly coastwise.
1977	189186	145315	43871	Petrol products approx. ¼ of total imports mostly coastwise; over 43000 tonnes casing; other notable items cement and metal manufactures. Crude minerals — over 50% of foreign imports.
1978	219290	175237	44053	39.89% petroleum products; 12.32% crude minerals. 10.54% casing; metal manufactures; chemicals
1979	238358	198913	39445	42.72% petroleum products; 13.13% casing; 12.14% crude minerals; 7.09% chemicals.
1980	253365	220396	32969	37.31% petroleum products; 15.3% casing; 15.27% crude minerals; liquid mud.

*1 *Coastwise: Within UK Continental Shelf Area including from rigs, production platforms lay and derrick barges.*
*2 *Foreign: Outwith UK Continental Shelf.*
*3 *No separation of figures was made before 1972.*

1a

IMPORTS FROM RIGS TO LERWICK HARBOUR
(since 1974 when separate figures first recorded)

Year	Total (metric tonnes)	Main Items
1974	4221	Empty containers
1975	3366	Empty containers and casing
1976	3765	Empty containers, casing, metal manufactures, (anchors, etc.)
1977	6132	Metal manufactures — over 40% casing, empty containers
1978	6557	Metal manufactures 57.83%, empty containers 17.11%, casing
1979	5179	Metal manufactures 44.27%; casing 20.39%; empty containers
1980	9135	Liquid mud 28.81%; metal manufactures 24.46%; empty containers, casing

COMMENT: 1979 figures reflected a downturn in activity offshore; 1980 figures illustrate growing importance of liquid mud plants in Lerwick.

1b

IMPORTS FROM PRODUCTION PLATFORMS TO LERWICK HARBOUR

Year	Total (metric tonnes)	Main Items
1975	105	80 tonnes empty containers
1976	978	36.8% empty containers; 35.7% metal manufactures, remainder rubbish
1977	5346	38.51% metal manufactures; 18.76% empty containers
1978	14722	32.99% metal manufactures; 22.22% liquid mud; empty containers, brine, casing
1979	23868	26.22% liquid mud; 19.97% metal manufactures, casing, empty containers, brine
1980	34945	33.95% liquid mud; 18.69% casing, metal manufactures, empties, brine

COMMENT: There were no production platforms offshore Shetland before 1975. Figures are expected to continue to rise as production phase offshore advances.

1c
IMPORTS FROM LAY AND DERRICK BARGES TO LERWICK HARBOUR

Year	Total (metric tonnes)	Main Items
1975	10587	75% concrete coated pipe. Remainder metal manufactures
1976	10169	Concrete coated pipe; metal manufactures
1977	3450	46.02% concrete coated pipe, 32.11% metal manufacture. Decrease in figures reflects decline in pipe laying
1978	17509	69.16% concrete coated pipe; water, bunkers
1979	4387	46.04% empty containers; 26.51% concrete coated pipe
1980	2793	64.76% empty containers; metal manufactures

COMMENT: Imports from lay and derrick barges are expected to increase, once new pipelaying programme commences.

OIL RELATED EXPORTS*¹ FROM LERWICK HARBOUR

Year	Total (metric tonnes)	Coastwise*² (metric tonnes)	Foreign (metric tonnes)	Main Items
1971	none	—	—	
1972	2996	2996	—	Barytes and water to rigs
1973	6110	6110	—	Barytes and water to rigs
1974	52814	52724	90	Barytes, water, chemicals and helicopter fuel to rigs
1975	211607	210803	714	33.08% concrete coated pipe; 32.13% water; 16.54% barytes; bunkers, cement, chemicals
1976	317884	314368	3516	22.02% water; 9.75% bunkers; barytes; concrete coated pipe; cement and chemicals
1977	200557	200469	88	25.9% bunkers; 25.5% water; crude minerals, casing
1978	265475	264229	1246	23.41% water; 19.69% bunkers; 13.02% metal manufactures; 12.5% crude minerals; casing, concrete coated pipe
1979	321518	321126	392	28.61% fresh water; 27.1% bunkers; 10.9% crude minerals; 9.37% casing, chemicals
1980	378919	373604	5315	31.86% fresh water; 20.22% bunkers; 15.1% casing, liquid mud, brine

NOTES:

*1 Exports here noted include exports to rigs, production platforms, lay and derrick barges.

*2 Coastwise means within UK Continental Shelf area.

2a
EXPORTS FROM LERWICK HARBOUR TO RIGS

Year	Total (metric tonnes)	Main Items
1974	52127	55.63% fresh water; 21.1% barytes; chemicals
1975	100119	13.98% barytes, helicopter fuel and bunkers; fresh water, chemicals, cement
1976	94216	50.94% water; 18.04% bunkers; barytes, chemicals, casing, cement, bentonite
1977	115547	38.98% fresh water; 22.71% bunkers, 19.08% crude minerals, casing, cement
1978	81682	36.51% fresh water; 27.5% bunkers; 18.01% crude minerals; chemicals, casing, cement, etc.
1979	42928	37.52% fresh water; 17.53% bunkers, crude minerals, casing liquid mud, brine
1980	74105	37.25% fresh water; 19.64% bunkers; crude minerals, casing, brine, liquid mud, cement

COMMENT: 1979 figures reflected downturn in exploration.

2b
EXPORTS FROM LERWICK HARBOUR TO PRODUCTION PLATFORMS

Year	Total (metric tonnes)	Main Items
1975	931	Metal manufactures, barytes, helicopter fuel
1976	30782	40.6% water; 25.98% bunkers; metal manufactures, barytes, cement, bentonite, containers
1977	57706	44.61% bunkers; water, metal manufactures, crude minerals
1978	152965	33.14% bunkers; 19.55% fresh water; crude minerals, casing, metal manufactures, chemicals, brine, petrol products, mud, cement
1979	252295	29.95% bunkers; 29.63% fresh water; crude minerals, casing; chemicals, metal manufactures, brine, mud, cement
1980	280565	32.64% fresh water; 21.67% bunkers; 13.4% casing; crude minerals, liquid mud, brine, chemicals, metal manufactures

COMMENT: Tonnages continue to rise as developments progress offshore.

EXPORTS FROM LERWICK HARBOUR TO LAY AND DERRICK BARGES

Year	Total (metric tonnes)	Main Items
1975	103886	67.38% concrete coated pipe; 20.58% crude minerals (crushed rock for pipe bed); water
1976	179331	83.89% concrete coated pipe; water, crude minerals, bunkers
1977	13037	50.24% concrete coated pipe; water, bunkers, metal manufactures
1978	17509	69.16% concrete coated pipe; water, bunkers
1979	15091	31.78% concrete coated pipe; 26.92% bunkers; 26.05% petrol products; water
1980	6630	47.93% petrol products; 23.57% fresh water; 18.76% bunkers

COMMENT: 1975 and 1976 figures reflect major pipelaying activity. Projected pipeline activity will probably cause an upturn in export figures in the next five years.

APPENDIX TO CHAPTER THREE
ORGANIGRAM 1
ORGANIGRAM OF COMPANY STRUCTURE OF NORSCOT OIL SERVICES LIMITED

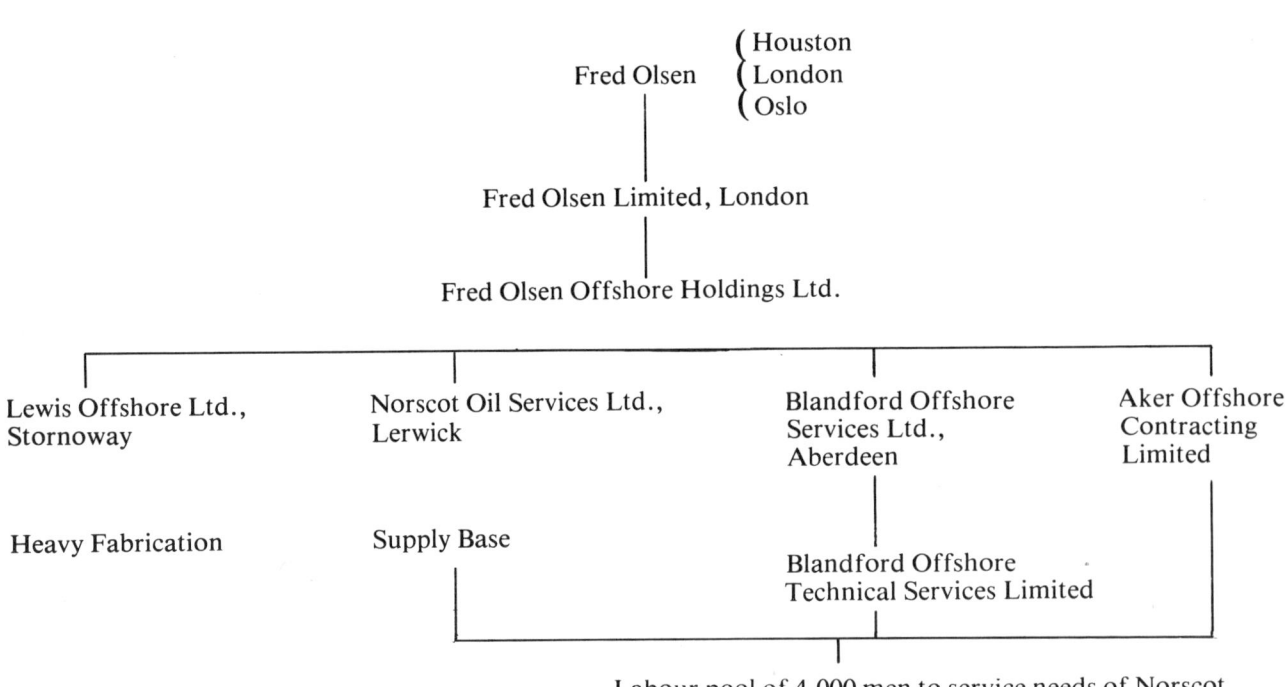

CLIENT COMPANIES AT NORSCOT BASE (as at January 1981)

Company	Function
Aberdeen Barytes Company	Barytes mill
Baroid (UK) Limited	Provision of oil based drilling mud
BP Petroleum Development (Sullom Voe) Limited	Administration for Sullom Voe Terminal
BP Petroleum Development Ltd. Operations	Induction and training
BP Petroleum Development Ltd. (Ninian Pipeline Operations)	Equipment storage for pipeline servicing
British Engine Insurance Co. Ltd.	Non destructive testing and inspection
British Oxygen Co. Ltd.	Supplier of industrial gases
B. W. Mud Ltd.	Supplier of drilling mud and of liquid brine as a drilling mud base for high pressure wells
CeBo (UK) Ltd.	Supplier of drilling muds and cement
Chevron Petroleum (UK) Ltd.	Transport and warehousing of materials for the Ninian Field
Commercial Catering (Scotland) Ltd.	Catering services
Dowell Schlumberger	Supplier of drilling muds
Dresser Magcobar	Supplier of drilling muds
Halliburton Manufacturing Ltd.	Supplier of drilling muds
Imco Manufacturing Services Ltd.	Supplier of drilling muds
International Drilling Fluids Ltd.	Supplier of drilling muds
J. M. Ironside Ltd.	Victualling services
Marinco (Marine Industrial Cleaning) Ltd.	Industrial cleaning and waste disposal
Milchem	Supplier of drilling muds
Oilfield Inspection Services Ltd.	Non destructive testing, inspection and heat treatment
S.G.B. Ltd.	Provision of scaffolding
Shet-Link	Road haulage and transport services
Staveley Electrotechnic Services Limited	Electronic care and maintenance services
Universal Rock Drilling	Drilling and blasting

OCEAN INCHCAPE LIMITED — BASE CLIENTS

Mud Companies

B. W. Muds	— Mud company
British Ceca	— Mud company
CeBo	— Mud company, 4 silos
Dresser Magcobar	— Mud company, 1 silo
Highland Muds and Chemicals	— Mud company, 1 silo

With Open Storage

Amoco, Conoco, Gulf, Mobil and Shell

Victualling

Magnus Smith — Cold store, ships chandlery

GREMISTA INDUSTRIAL ESTATE — *Summary of Current Situation (as at January 1981)*

TOTAL AREA — 13.47 acres
TOTAL No. OF SITES — 35

Serviced sites have been available since 1977. All except one site have been allocated, although not all companies have moved in.

Site Number	Company/Organisation	Activity
1	Viking Electrical	Electrical Installation/Repairs
* 3	Associated Tyre Specialists	Tyre Supply
° 4	Jim Johnson	Plant Hire - Trucks
° 5	HNP Engineers	Engineering Services
* 5a	Goudie & Mouat	Joiners
° 5b	Douglas Bentley	Video Studio
5c	Robert Elphinstone	Joiner
6 & 8	S. & J. D. Robertson	Fuel Trucks
* 7a & 7b	J. W. Smith	Cleaning Services
* 9	Malakoff Electrical & Electronics	Electrical and Electronic installation. Repair and maintenance work
10	Freefied Dairy	Milk Supply
*11a	C. H. Webster	Office Equipment Supply
11b	John Laurenson	Vehicle Maintenance
*14a	Shetland Freezer & Food Co.	Cold Store
*14b & 15a	R. E. Watt	Garage for Trucks
15b	Robertson & Johnson	Earth Moving Machinery
°16, 17, 18, 19, 20, 21, 22	Post Office	Engineering Division
*23	DITT Builders	Construction
*24	Plessey	Electronics and Installation. Repairs and maintenance
°25 & 26	J. W. Gray	Licensed Wholesalers
°27, 28, 29	North of Scotland Hydro Electric Board	Electricity supply equipment
*30	Schlumberger Inland Services	Oil Well Testing
30a	Vacant	

Footnote: * — *Site occupied, company installed*
° — *Under construction*

APPENDIX TO CHAPTER FOUR

COMPANIES OPERATING SCHEDULED FIXED WING AIRCRAFT SERVICES IN SHETLAND

AIR ECOSSE

This company began a scheduled service between Wick and Sumburgh in November 1979.

Head Office: Fairflight Charters, Biggin Hill, Kent.

Services Provided: Air charters; scheduled services between Aberdeen, Wick and Sumburgh; Aberdeen and Dundee; Dundee and Manchester; Wick and Glasgow.

Outwith Shetland: There are 11 aircraft positioned in Aberdeen including Bandeirantes, Twin Otters, Rockwells and Chieftain.

Aircraft used on Shetland Route: Bandeirante (up to 17 passenger capacity).

BRITISH AIRWAYS

Established in Shetland as British European Airways in 1946, became British Airways in September 1972, started operating as such on 1st April, 1974.

Head Office: Victoria Air Terminal, Buckingham Palace Road, London SW1.

Services Provided: Scheduled services from Sumburgh to Kirkwall, Inverness, Glasgow, Edinburgh, Birmingham, Manchester, London, Aberdeen. Also can provide charter flights from Sumburgh. Has extensive network of scheduled services throughout Britain and worldwide.

Aircraft used on Shetland routes: HS 748, capacity 44 passengers. Viscounts, capacity 60 to 70 passengers.

Fleet size: Approximately 180 fixed wing aircraft ranging from Concorde, Tristar, Trident, BAC 111, Boeing 747 and 737, Viscount to HS 748. Others on order.

LOGANAIR Ltd.

Formed in 1962, established in Shetland 1969.

Head Office: St. Andrews Drive, Glasgow Airport, Abbotsinch, Paisley, Renfrewshire, PA3 2TC. Bases at Aberdeen, Edinburgh, Kirkwall, Stornoway and Lerwick.

Shetland Office: Tingwall Airstrip, Gott, Shetland. Tel. Gott 246.

Services Provided: Scheduled and Charter services both within Shetland and between Shetland and the Mainland, Scottish Air Ambulance Service, Cargo and Aircraft Handling services.

Aircraft used on Shetland routes: Locally based, Britten-Norman Islander, 8 passenger capacity. *Glasgow/Aberdeen based,* De Havilland Twin Otter, 15-18 passenger capacity.

Fleet size: 7 Islanders, 4 Trislanders, 7 Twin Otters, 2 Bandeirantes, 2 SD-330s.

COMPANIES OPERATING FIXED WING AIR CHARTER SERVICES FROM SHETLAND

AIR ECOSSE

See above.

ALIDAIR Ltd.

Formed in January 1972, established from January 1974.

Head Office: East Midlands Airport, Castle Donington, Derbyshire.

Services Provided: Oil related charter work, mainly between Aberdeen/Sumburgh/Aberdeen. Contracts with BNOC, Conoco and Total.

Outwith Shetland: This company has a fully approved engineering facility, providing specialist maintenance for HS 125 and other aircraft.

Aircraft based on Shetland routes: Viscount 700's, 60 passenger capacity.

Fleet Size: 4 Viscount 700's; 2 Shorts SD 330's; 4 Shorts SD 360's on order for 1983 delivery.

DAN-AIR SERVICES Ltd.

Founded in 1953 as a subsidiary of Davies and Newman, the shipping brokers from whom the name is derived. Established in Shetland in August 1976.

Head Office: Bilbao House, 36/38 New Broad Street, London EC2M 1NH.

Services Provided: Oil rig crew change, and oil related construction support charter flights. Contracts for Shell and Unionoil, occasionally Chevron. Mainly Aberdeen/Sumburgh/Aberdeen; also for construction workers at Sullom Voe with Foster Wheeler Limited — mainly Sullom Voe/Glasgow/Sullom Voe flights, Dan-Air is the largest charter operator in Shetland.

Outwith Shetland: This company operates inclusive holiday tour flights, throughout Britain, and international flights throughout Europe. The main engineering base is at Lasham. Also scheduled service flights throughout Britain and Europe.

Fleet Size: (as at February 1981) 45 aircraft, including 19 HS 748's; 11 Boeing 727's; BAC 111's; 1 Boeing 737.

LOGANAIR Ltd.

(See listing under Scheduled Fixed Wing Aircraft Services).

HELICOPTER COMPANIES OPERATING OUT OF SUMBURGH AIRPORT
(as at February 1981)

BRISTOW HELICOPTER GROUP Ltd.

This originated from an independent company formed by Alan Bristow in 1951.

Head Office: Redhill Aerodrome, Redhill, Surrey. Bases also at Unst, Sumburgh, Aberdeen, Inverness, Teesside and Great Yarmouth. Also operations out of Blackpool and Haverford West (in Wales).

Services: The company has expanded into most areas of helicopter activity. These activities cover gas and oil rig support operations around the world, executive charter, search and rescue, civil and military helicopter pilot training and aerial crane work.

Helicopter Fleet Size: Through associate and subsidiary companies, the Bristow Helicopter Group Ltd. operate 193 helicopters and 12 fixed wing aircraft. This number includes 28 S61 N's; S58 ET's; 33 Bell 212's; 11 Wessex 60's; 11 Whirlwind WS 55 Series 3's; 1 Bell 204 B; 11 Bell 205-A-1's; 5 Alouette 111's; 25 Bell 47G Series; 9 Puma SA 330 J's; 1 Bolkow B0105; 12 Sikorsky S76A; 7 Hughes 500; 17 Astars; 20 Bell 206B and 2 Bell 206A Jetrangers.

On Order: 35 Puma SA 332L's, first deliveries expected 1982, at a cost of over £3 million each; 28 Sikorsky S76 A 'Spirits' at a cost of £1 million each — delivery continuing meantime.

Shetland Operations: These are carried out by Bristow Helicopters Ltd., a company which operates in the UK. There are 8 aircraft currently operating out of Shetland — 3 S61 N's and 1 Puma SA 330J at both Sumburgh and Unst. In addition there are Bell 212's operated by another Bristow Helicopter Group Company, British Executive Air Services Ltd.

BRITISH EXECUTIVE AIR SERVICES Ltd. (BEAS)

This is a subsidiary company of Bristow Helicopter Group Ltd.

Head Office: In early 1978 the Head Office of this company moved from Coventry Airport to Redhill Aerodrome, Redhill, Surrey.

Services: Gas and oil rig support flights, aerial and geographical surveys, aerial crane, executive charter, etc.

Helicopter Fleet Size: Amalgamated with Bristow Helicopter Group Ltd. (see above).

Shetland Operations: This company operates Bell 212 (14 passenger capacity) helicopters intermittently from Sumburgh.

BRITISH AIRWAYS HELICOPTERS Ltd.

Formed in 1964 as a subsidiary company of British Airways.

Head Office: Gatwick Airport south. Bases also at Aberdeen, Penzance, and Beccles, as well as Sumburgh.

Services: Offshore gas and oil rig and platform support, also search and rescue operations covering the North East coast of Scotland, including Shetland.

Helicopter Fleet Size: Total: 31 aircraft, 25 Sikorsky S61 N's; 2 Sikorsky S76's; 2 Bell 212's. 1 Jetranger; 1 Boeing Vertol Chinook (5 more on order).

Shetland Operations: These entail 11 S61 N's being based at Sumburgh.

NORTH SCOTTISH HELICOPTERS Ltd.

Head Office: Bourn Airfield, Bourn, Cambridge CB3 7TQ. This company also has bases in Aberdeen, Peterhead and Strubby (Lincolnshire).

Services: Management Aviation Ltd., and its wholly owned subsidiary North Scottish Helicopters Ltd., operate rig based helicopters offshore Shetland, using Bolkow 105 D's (4 passenger capacity), Scottish Express International handle their aircraft at Sumburgh airport, where the company built a new hangar costing approximately £250,000 in 1980.

Fleet Size: Total: 23 aircraft — 4 Sikorsky S61 N's; 4 Aerospatiale SA 365's; 4 Sikorsky S 76; 11 Bolkow 105D's. The company has an order, and expects to take delivery shortly, of another 4 Sikorsky S76's.

SUMMARY OF HELICOPTER AIRCRAFT TYPES CURRENTLY OPERATING IN SHETLAND

To date the S 61N and Aerospatiale Puma SA 330 J's have been significant on the Shetland scene, with the occasional Bell 212, Sikorsky S 58T and Bolkow 105D.

MAIN HELICOPTER TYPES OPERATING OFFSHORE SHETLAND

Type	Maximum No. of Seats	Speed (mph)	Range (miles)	Weight (lb)	Rotor Diameter
Bolkow 105	4 or 5	145	350	5070	32'3"
Sikorsky S-58T	16 to 18	130	300	13,000	56'
Sikorsky S61N	26 to 30	130	360	20,500	62'
Sikorsky S-76	12 to 14	145 to 167	460	10,000	44'
Bell 212	9 to 14	125	300	11,200	48'
Aerospatiale 330J Puma	19 to 21	155 to 160	340	16,315	49'3"
Boeing-Vetrol Commercial Chinook	44 to 46	153 to 165	575	47,000	60'

Notes: Range depends to some extent on the number of passengers carried. Ranges noted here presume that the maximum number of passengers are being carried.

For further explanations see 'The Helicopter in Civil Operations', by Captain Eric Brown, published by Granada, 1981.

AIRCRAFT HANDLING AND AGENCY COMPANIES OPERATING AT SUMBURGH

HAY & Co. (Lerwick) Ltd.

Old established Shetland company, created aircraft handling service at Sumburgh in June 1973.

Head Office: 106a Commercial Street, Lerwick.

Services Provided: Customs clearance, co-ordination of oil rig crew changes, general agency work. Agents for BNOC, Chevron, Conoco, Shell, Total, Unionoil at February 1981. Ad hoc agency for other oil companies as and when they are operating in the Shetland area.

Other operations in Shetland: At Lerwick, the Company acts as Lloyds Agents, operates also on coastal trade for the UK, and trades with Scandanavia, amongst other interests (see Chapter 3); also has an office at Sullom Voe where it conducts tanker agency work.

HUDSONS OFFSHORE

Does agency work at Sumburgh where it has an office (see Chapter 3).

NORSCOT OIL SERVICES Ltd.

Has done agency work at Sumburgh since end 1976/early 1977; mainly freight forwarding, and occasional work for exploration rigs in the vicinity (see Chapter 3, for further information on Norscot).

SERVISAIR

Established at Sumburgh Airport in January 1976.

Head Office: Barrington House, Heyes Lane, Alderley Edge, Cheshire SK9 7NE.

Services Provided: Freight agents; ticket agents; baggage handling contract for Dan-Air. Agents for Norwegian companies Norfly and Partnair which fly to Haugesund and Oslo; sells tickets for Air UK in Shetland for those wishing to travel from Aberdeen or other UK airports by Air UK.

Outwith Shetland: Offices at Belfast, Blackpool, Bristol, Cardiff, Chester, Derby, Edinburgh, Glasgow, Guernsey, Isle of Man, Jersey, Leeds, Liverpool, Manchester, Newcastle, Stansted, Swansea, Teeside.

SCOTTISH EXPRESS INTERNATIONAL

Established at Sumburgh in February 1976.

Head Office: 11 Alloway Place, Ayr KA7 2BS.

Services Provided: International aircargo brokerage (IATA members). Aircraft handling, charter brokers, crew change agency, customs brokerage (members of the Institute of Freight Forwarders Ltd.).

Outwith Shetland: Branches at Aberdeen, Edinburgh, Glasgow, Inverness, London, Manchester, New York, Newcastle. Also has shipping interests.

ANCILLARY SERVICES AT SUMBURGH AIRPORT

CROWN HOUSE ENGINEERING Ltd.

Crown House Engineering Ltd. have the maintenance contract for Sumburgh Airport, having been responsible for all maintenance work carried out at the airport since October 1979, and at present have a contract to continue this service until March 1983. To do this work they have nine operatives, three of whom are locally resident with their families. The remaining maintenance staff includes two locally resident secretaries, and two members of Ian Hunter & Partners, one of whom is locally resident.

This company commenced working at Sumburgh in mid-1978, and was involved in the installation of power cables from the existing Virkie side of the airport to the new Terminal site at Wils Ness. Thereafter they were involved in the installation work at the new Terminal, which commenced in September 1978, and was completed in May 1979. During that contract the company was responsible for the installation of all heating and ventilation systems, boilerhouse, chimneys, control systems, public address, flight monitoring and fire alarm systems, kitchen and lift installations. The total workforce, including subcontractors, peaked at approximately eighty operatives, who worked 12 hours per day, seven days per week throughout the contract. Following that work, the company was involved with installation of air conditioning plant, heating and ventilation systems, lighting, power, and standby generators for the new Visual Control Room complex.

This company is part of the Crown House Group, and accounts for about 70% of the group's total annual turnover, which is about £150 million. Crown House Engineering emerged from the amalgamation of Wheeler Crittall Berry Ltd. and Furse Electrical Installations Ltd. in 1973. It has 20 branches giving nation wide coverage, and is probably the largest electrical contractor in the UK, having 6000 employees.

WHIRLYBIRD SERVICES Ltd.

Established at Sumburgh since early 1977.

Head Office: Faucausie Grandhome, Aberdeen.

Services Provided: Survival clothing and equipment for helicopter passengers.

Outwith Shetland: Operates out of Aberdeen.

Earlybird Services which are a part of the Whirlybird Group, provide catering such as vending machines for hot drinks at Sumburgh — also at offices at Sullom Voe and Lerwick Swimming Pool.

WILS NESS HELIPORT

Design and Construction Team:

Main Contractors

Company	Function
Ian Hunter & Partners (in association with Sir Frederick Snow & Partners)	Consulting Structural Engineers
Ian Hunter & Partners	Consulting Mechanical and Electrical Service Engineers
G.R.M. Kennedy & Partners	Architects and Master Planners
Ian Peters & Partners	Project co-ordinator and Quantity Surveyors
Property Services Agency	Consulting Civil Engineers

Main Sub-Contractors

Company	Function
G. Percy Trentham Conder Scotland Ltd. Crown House Engineering Ltd.	Building Works
Costain-ARC Joint Venture Ltd.	Civil Engineering Works
Holiday Hall & Co. Ltd.	

SUMBURGH AIRPORT COMMITTEES

1. **Sumburgh Airport Consultative Committee** (meets as required)

   The Committee consists of:

   Four Shetland Islands Councillors;

   One Dunrossness Community Councillor

   CAA representatives, consisting of the Controller for Scotland and the Airport Manager;

   Representatives from helicopter and fixed wing operators at the Airport.

   The terms of reference of this Committee were agreed at their first meeting and are as follows:

   (i)   To stimulate the interest of the local population in the activities of the Airport;

   (ii) To consider any questions in connection with the airport as they affect the users and the community;

   (iii) To make suggestions to the Airport Manager on any matters connected with the administration of the airport which could further the interests of the users and communities;

   (iv) To advise the Airport Manager on any matters which may be referred to the Committee.

2. **Flight Operations Committee** (meets every two months)

   Representatives: All airlines;

   Scope: Flying operations;
                      Procedure and Equipment;
                      Ground Manoeuvring.

3. **Ground Operations Committee** (meets every two months)

   Representatives: All airlines;
                      Handling Agencies and Companies;
                      Police.

4. **Facilities Committee** (meets every two months)

   Representatives: All concessionaires;

   Scope: Provision of facilities, e.g. banking, catering, cleaning, taxis, etc.

5. **Security Committee** (meets every three months)

   Representatives: Members elected from Flight and Ground Operations Committees;
                      Balpa (British Airline Pilots Association);
                      Police.

   Scope: Airport security.

APPENDIX TO CHAPTER FIVE

ORGANISATIONS CONNECTED WITH THE SULLOM VOE ASSOCIATION

1. THE SHETLAND OIL TERMINAL ENVIRONMENTAL ADVISORY GROUP (SOTEAG)

Environmental assessment of the impact of oil-related developments at Sullom Voe was carried out until May 1976 by the Sullom Voe Environmental Advisory Group (SVEAG). At the end of this time, SVEAG produced a document on the environmental impact assessment of the Oil Terminal. The SVEAG was dissolved in 1976, as it was recognised that the composition and structure of the Organisation needed revision, if such a type of environmental group was to be effective in the operational phase of the Sullom Voe Terminal.

A new body, the Shetland Oil Terminal Environmental Advisory Group (SOTEAG) was therefore proposed, with the following representation:

	Organisation	No. of Members
1.	Shetland Islands Council	2
2.	Oil Industry — one representative from Brent Pipeline System and one from Ninian Pipeline System*	2
3.	Natural Environmental Research Council	1
4.	Nature Conservancy Council	1
5.	Countryside Commission for Scotland	1
6.	Department of Agriculture & Fisheries for Scotland	1
7.	Industrial Pollution Inspectorate	1
8.	Scottish Economic Planning Department	1
9.	Health & Safety Executive	1
10.	Shetland Fisherman's Association	1
11.	Shetland Bird Club	1
12.	The Universities	2

The Shetland Islands Council representation equals that of the Oil Industry at all times in this organisation. Both Oil Industry and Shetland Islands Council members are ineligible to hold any office in this organisation.

The Chairman is an independent member nominated by the Sullom Voe Association, which also provides secretarial help. The SOTEAG reports to the Sullom Voe Association. The main function of the SOTEAG at the present time is monitoring and surveillance in respect of the Oil Terminal.

2. SULLOM VOE OIL SPILL ADVISORY COMMITTEE (SVOSAC)

Until early 1980 two separate committees sat as follows:

1. The Industry Oil Spill Advisory Committee who were concerned with matters pertaining to oil spills on land and particularly within the Terminal; and

---

In the event of an additional pipeline, an additional oil industry representative could be appointed.

2. The Sullom Voe Oil Spill Advisory Committee which sat to consider relevant matters with regard to oil spills on the water.

The membership of both bodies was virtually the same but in one instance the Chairmanship was by the Terminal Manager and in the second by the Director of Ports and Harbours or Director of Construction, SIC. Discussions had been held for some time prior to this date regarding the merging of the two committees but it was felt inappropriate to undertake this exercise during the prolonged investigations which were taking place regarding the 'Esso Bernicia' incident* and only after the separate committees had made their reports was it considered appropriate to proceed with the merger of the separate committees. The new SVOSAC Committee, which reports direct to the SVA Limited consists of the following membership:

  Director of Construction, Shetland Islands Council, SIC
  Terminal Manager, Sullom Voe Terminal, SVT
  Director of Ports and Harbours, SIC
  Pollution Control Superintendent, SVT
  Oil Pollution Officer, SIC
  Ninian System Representative, SVT
  Director of Protective Services & Housing, SIC
  Brent System Representative, SVT
  Environmental Officer, SVT
  S. & J. D. Robertson Ltd., Shetland
  Shetland Towage Ltd., Shetland
(X) BP Environmental Control Centre, London
(X) Shell International Marine, London
(X) Esso Pet. Co. Ltd., Abingdon
(+) Senior Nautical Surveyor, Department of Trade

Members marked (X) will attend once per year but one such member, by arrangement, will attend each meeting. They may also attend, individually, any meeting at which items of specific interest to them are scheduled for discussion. They will, however, all receive the Minutes of each Meeting.

Member marked (+) will attend once per year but will receive the Minutes of each meeting.

Chairman

The Chairman will be elected either from the SIC or Industry on a rotating annual basis. The Chairman will hold office for at least one year, but no longer than two years, and in his absence the succeeding Chairman will chair the Meeting.

Terms of Reference for New SVOSAC Committee

1. To review policies and develop procedures, strategies and clean-up techniques.

2. To review preventive measures in the Terminal and Harbour Area and ensure that adequate plans exist for areas covered by the Sullom Voe Harbour Oil Spill Plan, the Terminal Oil Spill Plan and the Landward Sections of the Ninian and Brent Pipelines Oil Spill Plan, or such plans that may supersede any of these plans.

3. To periodically review and up-date all oil spill plans.

This incident occurred when there was a spill of over 1100 tonnes of bunker 'C' oil, at the end of 1978, from the tanker 'Esso Bernicia'. See Appendix to Chapter 6 'Oil Spills in Sullom Voe, since Terminal Operations Began'.

4. To recommend emergency equipment and materials, taking into account new developments in clean-up techniques. To recommend expenditure for such equipment.

5. To inter-relate items 1 to 4 with environmental advice from SOTEAG when necessary.

6. To ensure that equipment is properly stored and maintained.

7. To ensure that sufficient training programmes and exercises are carried out.

8. To monitor and investigate all incidents/spillages within SVOSAC's jurisdiction.

9. To take note of major pollution incidents world wide and to ensure that representatives are able to attend any such incidents where information/observation of equipment, etc., may be to SVOSAC's advantage.

10. To provide a Forum for the discussion of other oil pollution problems in the Shetland Area.

11. The Chairman to furnish a report of the Committee's activities to SVA Ltd. at least once per year.

3. SULLOM VOE OIL TERMINAL TECHNICAL WORKING GROUP

This Group was set up by the Sullom Voe Association in 1974 to consider all matters concerning the design and construction of the Oil Terminal.

Matters are remitted to the Technical Working Group which consists of about 12 personnel from the following:

1 **Shetland Islands Council:** The Council always has its oil consultant in attendance at these meetings. The Director of Ports and Harbours also attends when necessary;

2 & 3 **Shell and BP:** Each company usually sends about three representatives to these meetings;

4 & 5 **The Brent and Ninian Groups:** Each Group sends an observer.

4. THE JOINT EMPLOYMENT MONITORING GROUP (JEM)

No systematic collection or analysis of employment figures for the Sullom Voe area was in existence until about 1978.

The establishment of a second Construction Village at Toft gave rise to concern as to the possible implications of this development for the non oil-related industries. Shetland Islands Council, as Local Planning Authority, therefore made it a condition of the Toft Village Development that a liaison group, with representation from both the Oil Industry and the Council be established to monitor the employment situation.

The tasks of the Group are determined by the employment problems identified by it. These problems fall into the following categories:

(a) Employment problems of the non-oil industries. These problems are mainly related to losses of labour to the oil-related industries.

(b) Actual and potential problems connected with the Sullom Voi Oil Terminal labour.

(c) Actual and potential proportion of Shetlanders employed in the oil-related industry, and the main sources from which they are recruited.

(d) Unresolved training problems.

Tasks emanating from these identified problem areas include:

(i) Obtaining and regularly revising employment figures and problems connected with the traditional industries. Within this context attention will be paid to the different areas of Shetland in which these industries/firms may be situated as well as the industries/firms themselves.

(ii) Regular returns will be obtained from the relevant bodies in order to monitor and forecast both temporary and permanent employment in the Sullom Voe Area.

(iii) Training needs of the permanent work force will be studied.

The Joint Employment Monitoring Group reports its findings regularly to both the Sullom Voe Association and the Local Planning Authority.

It can be seen that whilst this Group is involved with the Sullom Voe area, its remit is wide ranging, and covers the Islands as a whole. Within this context, the JEM also liaises with a group that has been set up to monitor social change.

CONSTRUCTION CONTRACTORS AT SULLOM VOE
(as at August 1980)

Terms Used: Power Station and process plant are 'on site'; all other ancillary industrial activities are described as 'off-site'.

Notes: The listing shown is of main contractors currently involved in the construction of the Sullom Voe development and is not exhaustive. For ease of reference they are grouped broadly according to their responsibilities to one of the managing contractors on site. Contractors/suppliers who have been involved but are no longer active have been deleted from previous listings.

A section showing Shetland-based companies is also included with brief reference to the services they provide.

Company	Function
BP PETROLEUM DEVELOPMENT Ltd.	Constructor of the Terminal on behalf of the Sullom Voe Association.
Airwork Ltd.	Operational management of Scatsta airfield.
Capper Neill International Ltd.	LPG tankage contract.
Cape Contracts	Insulation.
Scaffolding (GB)	Scaffolding.
Commercial Catering (Scotland) Ltd.	Catering services outwith the Construction Site.
Constructors John Brown Ltd. (CJB)	Management and engineering contract for process facilities.
Dan-Air Services Ltd.	Air charter service.
Foster Wheeler Ltd.	Management and engineering contract for offsites and power station.

Foster Wheeler (GB) Ltd.	Common User Site Services. Include organisation of transportation and accommodation.
Franklin & Andrews	Quantity surveying services at Terminal site.
Grandmet (Shetland) Ltd.	Firth Village and accommodation ships' management and operation.
Motherwell Bridge Engineering Ltd.	Main tankage contractor for crude storage, tanks, ballast water and ancillary tankers.
J.D.&S. Tighe	Tank preparation and painting
Caleb Brett & Son Ltd.	Tank calibrating.
Chubb Fire Engineers Ltd.	Installing fire protection equipment.
Sea Truck Trading (UK) Ltd.	Provision of accommodation ship, Rangatira.
Stena Line A/B	Provision of accommodation ship, Stena Baltica.
Tarmac National	Management of services at Toft Construction Village.
Taylorplan Ltd.	Catering and cleaning services for Toft Construction Village.
Wilson Mason & Partners	Architectural services for Toft Village and permanent housing construction.
Wimpey Marine Ltd.	Construction Jetty operation.
FOSTER WHEELER Ltd.	Managing contractor, offsites and power station.
Balfour Kirkpatrick	Instrumentation cable jointing.
Butyl Rubber Company	Tank bund membranes.
C.B. Scaffolding	Scaffolding.
Dawson Keith Ltd.	Temporary site generators.
E.P.D.C.	Power station design consultants.
E.P.L. International Ltd.	Provision of aerial platforms.
Foster Wheeler Power Products	Installation and commissioning of boilers in power station.
Sir Alexander Gibb & Partners	Civil engineering consultants.
Harvey Cranes Ltd.	Craneage.
Holliday Hall	Power Station and offsites electrical and instrumentation installation.

L.J.K. Joint Venture (Lilley-JMJ-Keir)	Main civils contractor for roads, earth moving, peat disposal, site preparation, etc., bitumen and rock crushing services to entire site, concrete batching.
D.J.E. Seeding Services Ltd.	Grass seeding.
Edwards Reinforcements	Steelfixing.
Lawrie and Miller	Piperigging and welding.
Loyne (Belfast) Ltd.	Shotblasting, metal spraying, etc.
S.C.R. Construction	Shuttering.
Morceau	Fireproofing vessels and structures.
Joseph Nadin & Co. Ltd.	Power station and offsites insulation.
Norscot Oil Services Ltd.	Main flare fabrication works, other mechanical works.
Orbworld Ltd.	Instrumentation.
Plettac (UK) Ltd.	Temporary site structures.
Quality Inspection Services	Non-destructive testing, mainly X-ray.
Rigging International Ltd.	Main flare stack erection.
J.D.&S. Tighe	Preparation and painting for vessels, pipework and equipment.
Wimpey M.E.&C.	Offsites mechanical — erection of pipework and equipment.
Freeman Morrison Ltd.	Cladding.
CJB	Managing contractor, process facilities.

Much of the structural steelwork was prefabricated on the mainland and shipped to Sullom Voe for erection, complete with all vessels and pipework already installed.

N.G. Bailey	Electrical and instrumentation installation.
Cape Contracts	Insulation.
C.B. Scaffolding	Scaffolding.
Elliott Service Co. Ltd.	Specialist machinery alignment.
L.J.K. 3	Civils contractor.
A.J McLeod	Joinery.
Edwards Reinforcements	Steel fixing.

Miller and Lawrie	Piperigging and welding.
Universal Drilling Co. Ltd.	Rock blasting.
Morceau Fire Protection Ltd.	Fireproofing.
Oilfield Inspection Services	Stress relieving and heat treatment.
William Press	Mechanical installation of pipework and vessels.
Plettac Ltd.	Temporary structures.
Sparrows	Crane hire.
Rigging International	PAU transportation.
Ruberoid Contracts	Metal cladding to buildings.
J.D.&S. Tighe	Painting.
FOSTER WHEELER (GB) Ltd.	Main contractor, common user site services.
Abird Superior	Temporary generators.
Beatwaste	Sewage disposal.
Eflo Treatment	Sewage treatment plant.
A.J. Eunson	Coach transport — Scatsta to Villages.
G.E.S. Maintenance	Electrical maintenance.
Harvey Plant	Craneage.
Heenan Environmental System	Incinerators.
Kelvin Catering	Site catering.
Kinross Plant	Site transport — provision and maintenance.
L.J.K.	Rock, concrete, etc.
Pye Telecommunications	Site radio and internal telephone systems
Securicor	Site security.
Shetland Cleaning Services	Office cleaning.

SHETLAND BASED COMPANIES (some already covered)

A.J. Eunson	Coach service from villages to airport.
G.E.S. Maintenance	Electrical maintenance contractors
Hay & Co. (Lerwick)	Shipping agents.
Malakoff Ltd.	General engineering and electrical works.
Marjon Ltd.	Plant hire.
Nicholsons of Brae	Plant hire.
Norscot Oil Services Ltd.	Mechanical works.
P&O Ferries	Shipping agents.
S.&J.D. Robertson	Oil products distribution.
R.W. Offshore Services	General supply.
J.&M. Shearer	Shipping agents.
Shetland Aggregates	Rock and aggregate supply.
Shetland Cleaning Services	Office cleaning.
Shetland Line	Shipping Agents.
Sullom Quarries Ltd.	Rock, aggregate and concrete blocks supplier.
Sutherland Transport	Haulage, sub-contractor to Wimpey Marine.

APPENDIX TO CHAPTER SIX

Oil Spills in Sullom Voe since Terminal Operations began (Nov. 1978)

Date	Amount	Type of Oil	Source
30 Dec 1978	1174 tonnes	Bunker 'C'	Ship
20 Apr 1979	No quantity estimated	Dirty ballast	Ship
25 May 1979	50 gals	Crude oil	Crude loading
1 Jul 1979	10 gals	Crude oil	Ship
3 Jul 1979	100 gals	Diesel	Ship
21 Aug 1979	1 tonne	Bunker 'C'	Ship
6 Nov 1979	3 gals	Crude oil	Crude loading
4 Dec 1979	No quantity estimated	Diesel	Mavis Grind quarry
11 Jan 1980	250 gals	Crude oil	Jetty
12 Aug 1980	2 gals	Gas oil	Tug jetty
3 Oct 1980	250 gals	Crude oil	Ship
16 Oct 1980	20 gals	Gas oil	Tug jetty
12 Nov 1980	1-1½ tonnes	Dirty ballast	Loading arm/jetty
28 Nov 1980	2-6 tonnes	Crude oil	Jetty 2

MAIN POLLUTION EQUIPMENT STOCK PROVIDED BY TERMINAL OPERATOR

Held at Sella Ness

Dispersant (Concentrate) — in bulk storage tank	120 tonnes
Dispersant (Solvent based) — in bulk storage tank	40 tonnes
Chemical Barrier — in drums	5 tonnes
Warren Spring Laboratory Deep Sea dispersant spraying kits (adapted for concentrate spraying)	3
Warren Spring Laboratory Inshore dispersant spraying kits (adapted for concentrate spraying)	4
Knapsack Hand Sprayers	10

Beach Spraying Equipment (Cooper Pegler AR30 Pumps, hoses, lances)	5
Pillow Tanks (500 gallon)	20
Pillow Tanks (250 gallon)	3
Vikoma Seapack (450 m Heavy Duty Sea Boom) — with trailer	3
MARCO Boom Recovery Unit	2
Skimmex Boom (Medium weight) with anchors and ropes	4000 ft
Skimmex Boom (Light weight) with anchors and ropes	2000 ft
Komara Miniskimmers — with hydraulic power unit and pump	5
Slurp Skimmers with pump	2
Dunlop Dracone (45 m^3)	1
ORI Flexitank (13 m^3)	1
Collapsible Containers (1 m^3)	4
Atlanta Pumps with suction and delivery hoses	4
Spate Pumps (3″) with suction and delivery hoses	2
Mono Pump with suction and delivery hose	1
Absorbent Material — Drizit pillows	75 bales
— Conwed Sorbent Blankets	12 rolls
Simplon Portable Lighting Units with generator	4
Allday Aluminium 12 m Landing Craft, 'Voeclean'	1
Avon Inflatable Dinghy "Zodiac" with outboard engines	2
Road Trailer	2
Hand Pumps	3
Miscellaneous Equipment (spades, rakes, nets, plastic bags, etc.)	
Tool Kits	
Miscellaneous Spares	
Goggles and Protective Clothing	

Additional Equipment

Argocats Rough Terrain Vehicles	4
Trailers for above	2
Dispersant tanker (800 gallon capacity)	1
Jetvac Unit	

Held by SIC at Small Boat Harbour

Three Mooring Boats, each fitted with WSL Inshore Kit and tankage holding
    Dispersant Concentrate stock of 80 gallons
        'Sullom A' (Crew 1 + 2)
        'Sullom B' (Crew 1 + 2)
        'Sullom C' (Crew 1 + 2)
One General purpose launch 'Sullom Mareel', dispersant concentrate 250 gallons (Crew 1 + 2)
One Pilot Cutter 'Sullom Spindrift', dispersant concentrate 250 gallons (Crew 1 + 2)

Held by Shetland Towage Ltd.

Three tugs each fitted with WSL Deepsea Spraying Kit and tankage holding Dispersant Concentrate stock of 11.1 tonnes.

Equipment available from SIC (Director of Construction)
(Owned or hired equipment)
Selection of:
    Dump Trucks
    Tipper Trucks
    Skip Trucks
    Excavators
    Bulldozers
    JCBs
    Tractors
    Low Loader
    Cranes (6 and 25 tons)
    Gulley suckers

**Companies Related to the Sullom Voe Project
who were Employed by the Shetland Islands Council**

Byard Kenwest	Mainland suppliers of steel modular piles for the crude oil jetties
Christiani & Nielsen	Contracted to build the crude oil jetties at Calback Ness
Decca	Provided the harbour radar
George Dew & Co. Ltd.	Contracted to build the tug jetty at Sella Ness
Peter Fraenkel & Partners	Consulting engineers to the Council. Still working on the project at February 1981
Grampian Technical Services	Direct Shetland Islands Council contractors for supply of VHF equipment for the port. Still working on the project at February 1981
Jones & Healy Marine Ltd.	Nominated sub-contractors for the supply and installation of berthing monitoring systems for jetties 1 & 2. Main contractors jetties 1 & 2 Berthing Monitoring Systems. Still working on the project in February 1981
Knockbreda Builders Ltd.	Building contractors for the Port Administration Offices at Sella Ness
LJK	Tug and small craft jetty — Phase III
Mampaey	Provision of mooring hooks, capstans, etc., for jetties 1 & 2. Was a nominated sub-contractor for Christiani & Nielsen
Non-Corrosive Metal Products Ltd.	Access gangway for crude oil jetty No. 1 & 2. Nominated sub-contractor offsite — only on site to erect complete product
A. M. Robertson & Associates	Site agents and civil engineers to Stone-Platt Crawley Ltd.
W. H. Smith & Co.	Mainland suppliers of fabricated and structural steelwork. Direct sub-contractors to Christiani & Nielsen
Seebeck Aktein-Gesellschaft Weser	Mooring equipment for jetties 3 & 4. Sub-contracted to Christiani & Nielsen
Stone—Platt Crawley Ltd.	Principle contractors for the port navigation aids
Verhoef Aluminium Scheepsvouw Industrie of Allsmeer	Access gangways for jetties 3 & 4. Nominated sub-contractor to Christiani & Nielsen. Fabricated offsite — only erecting on site
Vreedestien	Provision of rubber fenders for crude oil jetties 1, 2, 3 & 4. Nominated sub-contractor to Christiani & Nielsen

The Zetland Harbour Advisory Committee

This is a statutory Committee, created by the Zetland County Council Act, 1974 (see Section 68). This Committee, consisting of up to 12 members who are appointed by the Secretary of State for Scotland, was constituted in 1980.

The Committee's functions are to advise and assist the Council in the exercise and discharge of its powers, duties, functions, and obligations, so far as they relate to the management, conservancy, control and development of the coastal area, and the maintenance, operation and improvement of port and harbour services and facilities in and in the vicinity of a harbour area, and upon any matter relevant to the foregoing matters which may from time to time be referred to the Committee by the Council.

The membership of the Committee is as follows:—

(a) one shall be a member of the Council;

(b) one shall be an officer of the Council;

(c) of the remaining members to be appointed, none of whom shall be a member or officer of the Council:—

    (i) at least three shall be persons appearing to the Secretary of State to be representative of those engaged in the fishing industry, using a harbour area (including at least one representative of each of the Shetland Fishermen's Association and the Shetland Fishmerchants' Association);

    (ii) at least one shall be a person appearing to the Secretary of State to be representative of the interests of those engaged in the hydrocarbon oil industry;

    (iii) one may be a person appearing to the Secretary of State to represent amenity interests;

    (iv) the remainder shall be persons representative of those using or providing port services or facilities in a harbour area.

APPENDIX TO CHAPTER EIGHT
POPULATION DISTRIBUTION BY AGE IN SHETLAND AND SCOTLAND 1961, 1966, 1971 and 1979

	1961		1966		1971		1979	
	Shetland	Scotland	Shetland	Scotland	Shetland	Scotland	Shetland	Scotland
MALES (%)								
0-14	24.6	27.6	24.1	27.5	24.3	27.6	26.6	23.8
15-19	6.7	7.5	7.4	8.6	6.9	7.9	7.1	9.2
20-24	5.9	6.4	6.4	6.5	7.2	7.8	7.1	8.2
25-29	5.6	6.5	6.2	6.1	6.7	6.3	8.9	7.2
30-34	6.1	6.6	5.5	6.1	6.0	5.9	9.7	7.2
35-39	6.8	6.9	5.7	6.3	5.7	5.9	7.1	6.0
40-44	5.6	6.2	6.9	6.5	5.6	6.0	5.3	5.8
45-49	6.9	6.6	5.3	5.9	6.7	6.2	4.4	5.7
50-54	6.9	6.6	6.6	6.3	4.8	5.6	4.4	5.7
55-59	6.7	5.9	6.2	6.0	6.3	5.7	4.4	5.8
60-64	5.8	4.6	6.0	5.2	5.8	5.3	3.5	4.4
65+	12.5	8.6	13.7	9.0	13.9	9.7	10.6	10.9
Total	8,510	2,482,734	8,305	2,478,750	8,430	2,514,620	11,294	2,489,461
FEMALES (%)								
0-14	20.7	24.2	21.2	24.2	22.0	24.3	25.0	21.0
15-19	6.3	6.9	6.1	7.7	6.2	7.1	6.5	8.2
20-24	4.9	6.4	5.4	6.1	5.8	7.1	7.4	7.3
25-29	5.3	6.2	5.1	5.9	6.0	5.8	8.3	6.6
30-34	5.2	6.3	5.1	5.8	5.3	5.6	8.3	6.6
35-39	5.8	6.6	5.3	6.1	5.2	5.7	5.5	5.6
40-44	5.7	6.2	5.9	6.4	5.0	5.9	4.6	5.6
45-49	6.5	6.6	5.8	6.0	5.8	6.1	4.6	5.6
50-54	7.4	6.6	6.4	6.4	5.6	5.8	4.6	5.7
55-59	7.2	6.1	7.2	6.3	6.2	6.0	4.6	6.1
60-64	6.3	5.4	7.1	5.7	6.9	5.9	3.7	5.0
65+	18.8	12.4	19.3	13.4	20.1	14.7	16.6	16.7
Total	9,302	2,696,610	8,940	2,689,460	8,895	2,714,340	10,817	2,677,539
TOTAL (%)								
0-14	22.5	25.9	22.6	25.8	23.1	25.9	24.9	22.4
15-19	6.5	7.2	6.7	8.1	6.6	7.5	7.2	8.7
20-24	5.3	6.4	5.9	6.3	6.5	7.5	7.2	7.7
25-29	5.5	6.3	5.7	6.0	6.3	6.1	8.6	6.9
30-34	5.6	6.4	5.3	6.0	5.6	5.7	9.0	6.9
35-39	6.3	6.7	5.5	6.2	5.5	5.8	6.3	5.8
40-44	5.6	6.2	6.3	6.4	5.3	6.0	5.4	5.7
45-49	6.7	6.6	5.6	5.9	6.2	6.2	4.5	5.7
50-54	7.2	6.6	6.5	6.4	5.2	5.7	4.5	5.7
55-59	6.9	6.0	6.7	6.2	6.3	5.9	4.5	6.0
60-64	6.1	5.0	6.6	5.4	6.3	5.6	3.6	4.7
65+	15.8	10.6	16.6	11.3	17.1	12.3	13.6	13.9
Total	17,812	5,179,344	17,245	5,168,210	17,325	5,228,960	22,111	5,167,000

Sources: Population Censuses 1961, 1966, 1971 and General Register Office for Scotland 1979 estimates.

Note: Percentages may not sum to 100 due to rounding

APPENDIX TO CHAPTER NINE

1. J. & M. SHEARER Ltd.

Background

This business was formed by the late Magnus Shearer (Convener 1929-35 and Provost 1940-46) plus his uncle the late James Shearer (hence J. & M.) in Lerwick in 1919 to engage in the herring curing industry, having at one time herring stations on the Island of Whalsay, at Collafirth in Northmavine and Cullivoe in Yell. The present site and private quay at Garthspool was acquired by Shearer's in 1926/27. The ownership of the private quay (there are only two in Lerwick Harbour — Hay's has the other) enables Shearer's to levy dues on all cargoes loaded and discharged there, although the Lerwick Harbour Trust can still levy vessel dues.

The family sold the business to the organisation H.B.P. Ltd. (Aberdeen) in 1970. H.B.P. formerly Herring By Products, whose parent company is Stord Bartz Industries of Norway, were the main marketers of the cured and salt herring in Britain at that time and oldest fish offal operator in Scotland.

Whilst herring curing was the core activity of the J. & M. Shearer business, the company had diversified to provide related services to the herring fishing fleet, for example, in 1945, the ice factory was built at Garthspool. The company also developed ships agency, stevedoring and engineering services for fishing vessels. In addition, the company provides consular services for Sweden and the Federal Republic of Germany.

Oil Related Business Developments

J. & M. Shearer required to diversify their business due to the downturn in the herring fishing. The best opportunities for expansion offered by the oil developments in Shetland were perceived as being in the engineering side of the business.

In 1975, Shearers therefore acquired the engineering business and workshops of T. W. Laurenson, servicing diesel engines and different types of marine engines. In addition, Shearers have the agency for several different types of marine engines, e.g. Caterpillar, Volvo and Perkins. (Mr T. W. Laurenson became a director of Shearers).

At the peak of the usage of Caterpillar machines at the Sullom Voe construction site (1976/7 to 1979) Caledonian, the main Scottish agent, set up a special workshop and supply depot adjacent to Shearers in Lerwick. This was closed by the end of 1979, when the main civil works at the Sullom Voe Terminal were completed. Shearers themselves, stuck to their traditional Caterpillar work with fishing boat engines and oil related Caterpillar engines.

In 1977, Shearers further expanded the engineering side of the business by forming a 50/50 joint venture company, Boyd and Shearer, with George Boyd of Glasgow, one of Scotland's largest engineering supplies companies. This allowed Shearers to operate on a much bigger scale and to stock a more diverse range of materials, supplying both engineering and construction contractors with everything from screws, nuts and bolts to doors.

In respect of the ships agency side of the business, Shearers have acted as agents for some cargo vessels at Sullom Voe, and have acted for Escombe McGrath in respect of tanker agency work.

Escombe McGrath, are however, a wholly owned subsidiary of P&O Ltd., and in November 1980, established a separate service in P&O's office at Sullom Voe for tanker agency work, leaving Shearers to continue their more traditional ships agency work with cargo ships and fishing vessels.

From 1976, Shearers quay has been used by Shetland Line Ltd., for the discharge of their vessels. This business developed further in 1978 when Shetland Line moved their main operation into Shearers warehouses.

Shearers contract the stevedoring necessary for this shipping operation which is another facet of the business since the Oil Era began.

Effects of Employment

The effects of the oil related expansion on the business of J. & M. Shearers has been to increase the full time employment staff five fold between 1970 to 1980 from 10 to 50 people. During the height of the herring season the company used to employ as casual labour a peak of perhaps 40 to 50 people, but it is the first time in the company's history it has been able to employ 50 full time workers all the year round.

2. JOHN LEASK & SON, Travel Agents

This family owned and run business, now Shetland's sole travel agency, begun in 1919 by operating as motor hirers. Motor hiring was the principal activity of the business until 1936 when airline agency work began.

The firm became agents for Aberdeen Airways De Havilland Dragon and Rapide aircraft which accommodated seven or eight passengers. This service was to continue throughout the Second World War. Aberdeen Airways eventually became Allied Airways and in 1937 they obtained the Royal Mail Contract for the letter post. This situation continued until 1947, when the airlines were nationalised and Allied Airways and Scottish Airways of Inverness amalgamated to become part of British European Airways. In 1972 British European Airways became British Airways and Leasks remain their agents to this day.

Leasks are also motor coach operators. Their first buses were 14 seater hooded charabancs in the 1920's. Leasks maintain this business activity to the present day, although the size and appearance of the buses has somewhat altered. This side of the business has increased from seven buses in 1972 to ten buses and two mini buses in 1980. The company has contract work transporting construction and catering workers to Sullom Voe. There has also been an increase in the number of taxis owned from six to nine. The 12 self-drive cars Leasks have, have not increased in number since 1972 as they decided not to expand that side of the business.

Since the Oil Era the Company has noticed many changes. Travel business has increased about five fold, and is no longer seasonal, people going on holiday all the year round. Before 1974, there was very little international business. Due to changes occurring, Leasks became IATA (International Air Transport Association) agents, able to provide service for all airlines worldwide.

It is not just incoming people connected with the oil business who have stimulated Leasks travel business; amongst local people there has been a very great increase in travelling.

Leasks' estimated oil related activity could account for 55 to 60 percent of their business, involving them in bus, taxi, and self-drive hires, and the movement of employees of oil related companies connected both with Sullom Voe and offshore operations by air, sea and rail (Leasks are British Rail agents). Now that the Sullom Voe Terminal is operational, they also provide services for tanker crews.

With the small but viable increase in population due to Sullom Voe operation staff families moving to Shetland, Leasks expect the areas of business not normally considered as oil related such as the package holiday and friends and relatives' market to stabilise.

The firm has increased its employment from 10 to 19 people since the Oil Era began, only one of these not being local (an Orcadian).

3. J. M. IRONSIDE, Wholesalers

The proprietor of this business, Mr J. M. Ironside, arrived in Shetland in 1970, and at first began in business as a butcher. By 1973/74, he sold his butchery business and established himself as a wholesaler in partnership with J. & P. Merchants of Aberdeen, mainly for food items. He moved into premises at Norscot Base, and operated from there for three years, until 1977.

In 1977 two major changes occurred which were to affect this company's future history:
1. Toft Construction Village was built.

2. Ironside joined a consortium of wholesalers known as the 'Spar' group.

By obtaining a wholesale contract to supply Toft Construction Village, Ironside was assured of a high sales turnover for more than four years. This gave him a base from which he could expand other aspects of his business especially providing a service for retail outlets.

By joining the Spar buying consortium, Ironside was to become one of a group of 14 wholesalers whose organisation covers the whole of Britain (except Orkney). This consortium have a total turnover of £600 million per annum and operates through a central purchasing office in London. An idea of the scale of business conducted is that the London office costs £3 million per annum to run, and in addition the Spar group spends £3 million per annum on advertising. Ironsides is the smallest of the wholesalers in this consortium.

With the rundown of construction at Sullom Voe, Ironside hopes to diversify and to concentrate more on the retail side of his business.

The Oil Era has generated sufficient business to give Ironside a sound basis on which to develop; by generating this business it allowed him to become a member of the Spar group which gives him access to centralised buying and helps him to be competitive. Had the Oil Era not happened, J. M. Ironside, Wholesalers, probably would not have either.

In 1980 Ironside was employing 13 staff.

4. MALAKOFF Ltd., North Ness, Lerwick.

The local area known as the North Ness has since about 1880, been the site of a fishing vessel repair yard and today it is the yard best known as "The Malakoff". Various owners have occupied the area and until 1973, it was run by the Johnston family who also have building interests in Lerwick.

In 1973, the business was acquired by the Lithgow Holding Company of Glasgow. Lithgow's involvement in Shetland began in 1969, when they were commissioned to make a design of vehicle ferries suitable for use in Shetland. They had a further involvement in 1972 when they were asked to prepare a study of the potential areas in Shetland for development of a major oil terminal, the areas being specified by the Shetland Islands Council.

The business of The Malakoff was to provide ship repair facilities to the local fishing fleet and to other users of the Lerwick Port. This policy has been maintained under the new owners, but the yard and the facilities of the yard have been improved to enable the company to meet the requirements of the major civil engineers and others during the construction of the Sullom Voe Terminal.

The involvement of Lithgows in Shetland took a further step in 1974 when the John Wood Group of Aberdeen purchased the interests of William Moore & Sons of Scalloway.

The Scalloway yard also has a history in ship repairing and for many years was run by Mr J. Moore, and his family before him. A decision was taken in 1974 by the main Board of the two companies* that a working arrangement between them, as a group rather than individual units, would enable them to provide an improved service to local industry. The management of this group today is in the hands of Shetlanders and with a firm policy of apprentice training and recruitment, the group now employ within the ship repair divisions a total of 75 people.

With developments taking place in Shetland, not only with the oil industry, but also with the new family of large fishing vessels, it has been necessary for the group to continually improve their facilities. Included in this is a new cradle for the Lerwick slipway to enable the yard to handle the larger purse net vessels, as a result of which Malakoff are now in the situation where only three of the newer vessels are beyond the capacity of the yard.

At present the Scalloway yard is undergoing a further development with the extension to the main jetty and slipway track. This will be completed by mid-1981 and whilst this development will not increase the capacity of the yard, it will enable them to handle and slip the vessels at normal tides.

The policy of the group is to extend where practical and to provide services to meet the growing age of technology and in 1977 a further division was developed on the Gremista Industrial Estate.

The Wood/Lithgow joint venture began in 1974.

This company is known as Malakoff Electronics and Electrical Ltd., and they can provide services to other local industry in disciplines associated with the electrical and electronic side of engineering. This activity extends to a contract with the Ports and Harbours complex at Sellaness and also to the oil terminal at Sullom Voe.

The Oil Era has been a challenge to this traditional industry but this challenge has shown that certain aspects of the requirements can be met within Shetland. Oil, whilst bringing problems, has also highlighted to local industries the need to move in the direction of service and quality and this will be the continuing policy of Wood Lithgow Ltd.

6. BOLTS MOTOR GARAGE Ltd.

This is a locally owned business which began in Lerwick in 1950. It has three main areas of business activity — vehicle rental, retailing fuel, and retailing vehicle parts.

By 1980 the company had operated vehicle contracts for seven years with all the major companies involved at Sullom Voe — both in the constructional and the operational phases. It was anticipated that the company would be affected by the rundown of construction activities at Sullom Voe, and that they might require to reduce their vehicle fleet by up to 15 percent, but there is currently no sign of this happening, the fleet actually having increased in size!

The company estimated 70 percent of its vehicle rental, 20 percent of the retail fuel and 35 percent of the retail parts business to be oil generated. From 1972 to 1980 the number of employees of Bolts Motor Garage grew from 18 to 59.

Bolts Motor Garage is part of a group of companies, three of which — Shetland Entertainment Co. Ltd., Northern Lights Investments Ltd., and Bolts Mossmorran Ltd. — have emerged with the Oil Era, in 1977, 1978 and 1980, respectively.

The Shetland Entertainment Co. has existed since 1959 but has only operated in its present form since 1977, when the Chapel House Restaurant was created from the former Planets dancehall. In 1980, this company employed 45 people having not only the Chapel House Restaurant but also the catering concession at the Jubilee Centre. There is also a rapidly expanding outside catering business which provides services to functions throughout the Islands.

The Northern Lights company owns the Brae Hotel which was opened in 1978. It employs 43 people of whom only six are local. The company has built four houses for staff accommodation and also provides accommodation for staff at the hotel, it having proved difficult to find local staff since the Oil Era began.

Bolts Mossmorran Ltd. is perhaps one of the most interesting business developments to result from the Oil Era. The company decided to utilise its Shetland experience in car and vehicle hire and bought a garage property in Cowdenbeath to help service the Mossmorran development. There is a fleet of vehicles — Land Rovers, vans and mini buses, as well as car hire and petrol outlet. The company employed eight people in 1980 with plans for expansion. None of these employees were Shetlanders all being local Fife people. The garage foreman was a Fife man employed in Shetland by the company who was moved back to Fife when this development began. This has been a reversal of the trend for outside owned companies coming into Shetland and employing Shetlanders — now a Shetland company has gone outside to do likewise!

The last link in the Bolts Chain, not mentioned so far is Sandwick Transport Ltd., a coach hire and taxi company set up in 1950. This company employs 15 people and has, like the others, expanded since the Oil Era, now operating more buses.

7. SHET-LINK TRANSPORT SERVICES Ltd.

This company was founded in February 1978, as a limited company with five shareholders, three of them being native Shetlanders. The company has its office at Norscot Base in Lerwick, and two of the Directors are based on the Scottish Mainland, one at Aberdeen, and the other at Perth.

The company provides a service both to the business community, and to the ordinary household, having three articulated units, and seven trailers for its industrial traffic and three vans for the domestic side of the business, which includes a door-to-door removal and storage service.

Since 1979, the company has rented a warehouse at the Gremista Industrial Site. The company estimated that approximately 50 percent of its business is directly oil generated. Oil related activities with which the company has been involved include carrying steel to Shetland for Norscot Oil Services, and transporting drilling mud and chemicals to the Scottish Mainland for the mud companies.

The company also imports food for the wholesalers in Shetland.

Shet-Link provides a service from Shetland to all points in Britain, and are about to extend their operations to include the continent of Europe.

At February 1981, the company employed ten people.

8. **MILLER CONSTRUCTION NORTHERN Ltd.**

This company has had a continuing involvement in Shetland since the early days of the Oil Era. Involvement in Shetland began midway through 1974 when the company were awarded the first phase of the Firth Construction Village buildings, by the Zetland County Council. Subsequent development at Firth was also awarded to Miller Construction, followed by the award of the Toft Construction Village contract by BP Petroleum Development Ltd.

As the developments continued apace in Shetland, Millers constructed a wide variety of buildings, employing over a thousand men (including sub-contractors) at the peak of operations in 1977. Completed work done by this company includes schools at Brae, Sound (Lerwick) and Mossbank; housing at Sumburgh, Strom, Upper and Mid Lea (at Firth) and Brae; offices, hostels, and hotels; a telephone exchange; sewage and water treatment works. By 1980 construction work undertaken by Millers in Shetland exceeded £50 million worth of contracts.

Miller's success, especially earlier on in the Oil Era, demonstrates that few Shetland companies were geared up at that time to cope with contracts of such magnitude. Further, it has always been a policy of Millers to employ local Shetland people where possible, and in 1981, they continue to do this. Presently, Millers are constructing the Engineering Services building at Sullom Voe Terminal, which it is intended to use at the opening of the Terminal on 9th May, 1981, by Her Majesty the Queen.

Miller Construction Northern Ltd. is the principal construction arm of the Edinburgh based Miller Group, who have wide ranging interests in the UK as well as business interests in the United States of America.

Glossary

API: American Petroleum Institute — an organisation incorporated in the United States of America (U.S.A.) which is, amongst other functions, the recognised authority for petroleum operational and equipment standards.

API Gravity: The arbitrary scale used by the American Petroleum Institute for assessing the quality of oil. The unit of measurement is in degrees. The American Petroleum Institute degrees is derived from the following formula:

$$API° = 141.5 \frac{\text{(Weight of 1 volume of water at 60° Fahrenheit)}}{\text{(Weight of 1 volume of oil at 60° Fahrenheit)}} - 131.5$$

or $\frac{141.5}{\text{Specific gravity of oil}} - 131.5$

For example, on the API scale, the density of water would be described as 10°API. The specific gravities of North Sea oils mostly tend to be lower than water giving them a high API rating — usually in the 35° to 45° range. Oils in this range are described as "light crude", which means they contain a higher proportion of light oils such as gasoline and kerosene. "Heavy crude" oil has a much lower API, for example, the 15° to 25° API range, and yields types of heavy fuel oils such as those used for bunkering ships and for other industrial purposes.

Associated Gas: Natural gas associated with oil accumulations, which may be dissolved in the oil under reservoir temperatures and pressures, and separated out when the oil is produced, or it may form a cap of free gas above the oil in the reservoir.

Atmospheric Pressure: The pressure of air at sea level. As a standard the pressure at which the mercury barometer stands at 760 millimetres or 30 inches (equivalent to 14.7 pounds per square inch).

Barrel (bbl): A common unit of measurement of liquids in the petroleum industry.

> 1 barrel = 42 U.S.A. gallons or 35 Imperial gallons or 159 litres (approx.), or 135 kilogrammes (approx.).
>
> For general purposes (average) calculations:
>
> 1 barrel = tonnes ÷ 7.4.
>
> 1 tonne = barrels x 0.13 tonnes.
>
> Barrels per day = tonnes per year x 0.02; tonnes per year = barrels per day x 50.
>
> For energy equivalent calculations:
>
> M^3 = barrels x 150; barrels = M^3 x 0.0067.

The number of barrels that can be yielded from one tonne of petroleum (bbls/tonne) depends on its specific gravity (SG). (For definition of specific gravity see Glossary). An illustration of this is seen below:

Specific gravity (SG) versus barrels/per tonne (bbls/tonne)

API	SG	bbls/tonne
25	0.90	7.0
30	0.88	7.2
35	0.85	7.4
40	0.82	7.6

Thus, for truly accurate calculations of tonnages of petroleum yielded from various reservoirs, one needs to know the Specific Gravity of the petroleum concerned, which tends to differ from field to field.

Barytes: Barium sulphate, a heavy mineral of high specific gravity used in drilling mud. It is mixed in powdered form to increase the density or weight of the mud and thus control pressure in an oil or gas well so as to prevent the well blowing out. It also helps lubricate the drilling bit.

Bentonite: The clay mineral montmorillonite, a magnesium- aluminium silicate used in drilling as a mud component, in greases, and in the refinery as a treating agent.

Bunkers: Any fuel oil or diesel fuel taken into the bunkers of ships.

Butane: (C_4H_{10}). A colourless hydrocarbon gas of boiling point 31° Fahrenheit, 0° Centigrade. Commercial butane is a mixture of gaseous paraffins, mainly normal-butane and iso-butane (both C_4H_{10}). When blended into gasoline in small quantities it improves volatility and octane number. Butane can be stored under pressure as a liquid at atmospheric temperatures and is widely used as bottled gas for small scale domestic cooking and heating, when pungent chemicals are added to butane for safety reasons.

Calorific Value: The amount of heat obtainable from the complete combustion of a unit weight of fuel. Calorific value is normally expressed as calories per gram or B.Th.U's per pound (lb), (a B.Th.U. or British Thermal Unit is the heat required to raise the temperature of 1 lb of water through 1 degree Fahrenheit).

Casing: Steel lining used to prevent caving in of the sides of a well, to exclude unwanted fluids, and to provide means for the control of well pressures and oil and gas production.

Cement/cementing of wells: Filling part of the space between the casing and bore hole wall with cement slurry. On hardening it keeps the casing in the hole stationary and prevents leakage from or to other strata that have been drilled through.

Concrete coated pipe: Submarine pipelines have to be coated in concrete to give added weight, to prevent corrosion and afford protection from ships' anchors, trawl boards, etc. This operation is done onshore and the coated pipe then transported by supply vessel or specialised hauling vessel to the lay barge. When the sections are finally welded on the barge, the breaks between the concrete coating on each individual joint are protected by a special cladding put on after the weld has been inspected. Land pipelines are usually coated with a bitumen or mastic substance.

Continental Shelf: The shallow, submerged 'platform', bordering and marking the structural edge of the continent. The use of the term 'shelf' is reserved for the seabed of a depth of 200 metres or less. Beyond the shelf is the continental slope and then the deep ocean.

Crude Oil: The oil production from an underground reservoir, after being freed of any gas which may have been dissolved in it under reservoir conditions, but before any other operation has been performed on it. In the oil industry, simply termed 'crude'.

Derrick barge: A crane barge used in the offshore construction industry, suitable for working in rough seas. These barges are equipped with heavy-lift cranes capable of lifting platform decks, modules and other large items of equipment.

Drillship: Free floating ship shaped vessel which is kept in position by multiple anchors or by dynamic positioning.

Drill string: The column of drill pipe and drill collars screwed together, at the end of which the drill-bit is screwed.

Drill pipe: The steel pipe used for carrying and rotating the drilling tools and for permitting the circulation of the lubricating mud.

Dynamic positioning: Method of maintaining a vessel or floating structure on location by means of an Automatic Station Keeping system (ASK). Information from sensors on the seabed and in the ships hull is fed automatically to the computer which converts the information into appropriate commands to the thrusters and/or variable pitch propellers which maintain the vessel on station. This system allows vessels to drill in deep water in which it would be inconvenient or impossible to anchor.

Ethane (C_2H_6): A colourless, odourless light gas commonly found in natural gas. (See also *gas*).

F.O.B.: Free On Board, i.e. the export price of crude when it has been loaded onto a tanker.

Flare stack: Pylon like structure through which flaring occurs.

Flaring: The means of safe disposal of hydrocarbons by burning. Offshore fluids produced in well tests are usually flared because of a lack of storage and processing facilities. Natural gas may also need to be flared if it is not injected back into the reservoir or if there is no means of transporting it to shore. At a shore terminal, such as Sullom Voe, flaring allows small quantitites of light gases which are uneconomic to process to be burnt off and also acts as a safety precaution should pressure build up unduly in the processing activity. Flaring is strictly controlled by the Department of Energy.

Flash point: The lowest temperature at which vapours arising from the oil will ignite momentarily (ie. flash) on application of a flame under specified conditions.

Floating roof: A special tank roof which floats upon the oil.

Gas: Lighter Hydrocarbon gases can appear on the surface from a producing reservoir in various forms and combinations. The four gases at the lowest end of the Carbon chain are the most commonly found, viz:

      Methane (CH_4)
      Ethane (C_2H_6)
      Propane (C_3H_8)
      Butane (C_4H_{10})

The next hydrocarbons in the carbon chain — pentane, hexane, heptane, etc. generally remain in stabilised crude oil.

Gas/Oil Ratio (G.O.R.): The quantity of gas produced with the oil from an oilfield, usually expressed as cubic feet per barrel of oil or as volumes of gas per volume of oil.

Gas gathering systems: See Appendix to Chapter One.

LPG — Liquefied Petroleum Gas: Mainly propane and butane, which can be liquefied by pressure alone (typically 6 to 7 atmospheres) although chilling helps lower vapour pressure.

Lay barge: Specialised barge or vessel used for laying submarine pipelines. It may be designed on the semi-submersible principle. 'Joints' (40 foot sections) of pipe (or in some cases double joints — 80 foot sections prewelded onshore) are welded together on the vessel and fed over the stern ramp or stinger as a continuous pipeline onto the seabed.

Liquefied Natural Gas (LNG): In the North Sea this is usually over 90% methane which has been liquefied for tanker transportation by refrigeration alone, to minus 161° Centigrade, with the liquid remaining at atmospheric pressure.

One tonne (approx. 16 bbls) of LNG regasifies to approximately 1400M^3. This is approximately equal to 1.3 tonnes or 9.5 barrels of oil.

Median Line: The line decided by International Agreement which marks the boundaries of Sectors belonging to individual states bordering the Continental Shelf, dividing the Continental Shelf sea area into areas of national jurisdiction for the purposes of the exploration for and exploitation of seabed minerals.

Methane (CH_4): A light odourless, flammable gas, of boiling point $-151.4°$ Centigrade. It is the chief constituent of natural gas. It is often produced by decaying plants in swamps and its presence can be misinterpreted as an indication of the presence of petroleum.

Module: The box or 'package' containing equipment for installation on a production platform. These modules weighing up to several thousand tonnes each, are constructed onshore and installed as self contained units on the deck of the offshore platform. Steel platforms are floated out horizontally and the modules have to be lifted on after the installation and piling of the platform using a crane (derrick) barge or Maintenance Construction Barge (MCB) of which there have been several offshore Shetland (see Lerwick Harbour Statistics in Appendix to Chapter Three). Normally each module serves a specific purpose eg. accommodation, compressor, generator, gas separation, etc.

Moonpool: The well or hole which gives direct access to the sea, in for instance, a drillship, diving support vessel or semi-submersible rig. In the case of ship-shaped vessels, the moonpool is usually amidships. On a drillship (eg. Petrel) it is through the moonpool, directly beneath the derrick, that the riser pipe extends to the seabed, and through which drilling is carried out. In purpose built diving support vessels the diving bell is also launched through a moonpool rather than over the side or stern of the vessel.

MSV: Multi purpose support vessel; whilst able to carry out support work for maintenance and construction an MSV is also equipped for firefighting, pollution control and oil-spill clean-up, emergency rescue and salvage. Ideally semi-submersible and dynamically positioned, such a vessel would normally service a group of offshore fields for example, working for a Safety Sector Club. (See Safety).

Mud (drilling): In the rotary drilling of wells it is essential to remove the drilled material or cuttings from the bore hole by circulating suitable pumpable fluids. These fluids can be water, seawater, or oil based, and contain finely ground minerals such as barytes, shales or special clays. It is pumped through the drill-string to the bottom of the bore-hole, whence it rises to the surface through the space between the drill string and the bore-hole wall.

NGL: Natural Gas Liquids are the heavier gases, ethane, propane and butane, possibly going as high as heptane (C7), which can occur as liquids.

Natural Gas: Gas issuing from the earth under pressure and often produced in association with crude petroleum. Natural gas is usually classified as "wet" or "dry" depending on whether the proportions of gasoline constituents which it contains are large or small.

Net cash flow: Cash income minus cash expenditure.

Nautical mile: One nautical mile = 6080 ft. One mile = 5280 ft.

P.A.U.: Pre Assembled Unit. In terms of the Sullom Voe Project this meant one of the 'modules' which were assembled on the UK mainland, and shipped up to the site, in order to ease construction.

P.A.R.'s: Pre Assembled Piperack. Usually pipework module for gas processing at the Sullom Voe Terminal.

Pig: Device forced through pipelines, usually by hydraulic pressure, to detect flaws, scrape off rust and scale or to mark an interface between two different products. It is also known as a sphere and can be self propelled carrying increasingly sophisticated instrumentation for recording pipeline parameters such as variations in diameter, ovality, obstructions, deformations, etc.

Pig receiving station: Equipment at receiving end of a pipeline in which the pig is recovered after leaving the pipeline and where the material pushed along the pipeline by the pig e.g. waxy deposits, rust, etc., is deposited.

Plugging a well: When a well is abandoned for any reason, either temporarily or permanently it must be sealed off to ensure that no escape of any substance can occur. This sealing off is generally accomplished by insertion of a plug of cement after using brine (see Chapter Three — Drilling mud and Cement Business).

Propane (C3H8): A colourless hydrocarbon gas of boiling point of $-43.7°$ Fahrenheit, useful for heating, metal cutting and flame welding purposes. It can be stored under pressure as a liquid at atmospheric temperatures but is more volatile than butane and higher pressures are required to keep it in liquid form. It is useful for both central heating and domestic cooking, having the advantage that under British climatic conditions, it does not freeze in winter.

Reserves: Amount of crude oil or gas expected to be recovered profitably from a given reservoir.

Sour crude: Crude oils containing a large amount of sulphur and sulphur compounds which produce undesirable corrosive sulphur compounds when refined.

SALM: Single anchor leg mooring (buoy). See also SBM.

SBM: Single buoy mooring used for loading oil into tankers in the open sea. Also sometimes called single point mooring (SPM). The principle is that the tanker can moor to load oil whatever the direction of wind or current and swing at its mooring to present the least resistance to the prevailing conditions.

Scf: Standard cubic feet. Gas production is measured in scf ie. cubic feet under standard conditions of 60 degrees Centigrade and atmospheric pressure. The metric equivalent is NM^3 or normal cubic metres at 0 degrees Centigrade and atmospheric pressure. One NM^3 is approximately equal to 37.3 scf. MScf = one thousand standard cubic feet. MMScf = one million standard cubic feet. Scfd = Standard cubic feet per day.

Semi-submersible: Any vessel — drilling units, MSV, lay or crane barge, which is supported on pontoons submerged sufficiently beneath the surface of the sea to be below the level where wave motion is greatest, rather than having a conventional hull formation as in ship-shaped vessels.

Show: The term used to describe an indication of the presence of oil or gas, whether or not in commercial quantities.

Sour gas: Hydrocarbon gas containing undesirable sulphur compounds sulphuretted hydrogen and methyl mercapton.

SPAR: A type of single buoy mooring (SBM) developed by Shell, incorporating storage facilities, so that in the event of weather conditions temporarily preventing tanker loading, production need not be shut off. The letters SPAR stand for 'seagoing platform for acoustical research', the original design, on a smaller scale, being used for that purpose. This design was subsequently modified by Shell for use on the Brent Field.

Specific gravity (SG): The ratio of the weight of a given volume of substance to the weight of an equal volume of water at the same specified temperatures, normally 60° Fahrenheit. (For uses of SG see **API Gravity** and **Barrel** definitions).

Spud or Spud in: To commence drilling operations by 'making a hole'.

Standby vessel: A small vessel, often a converted trawler, chartered for the sole purpose of remaining on location as near as possible to the offshore installation in case of accidents. There is a statutory obligation on every offshore operator, whether of a fixed or mobile structure, to have a standby vessel in attendance, and these vessels must meet with certain standards, for example, the number of survivors they can take. Standby vessels should not be confused with supply or service vessels, which are used to carry supplies to offshore installations and are often larger more versatile purpose built vessels, capable of fulfilling other functions such as anchor handling, towing, etc., and having open deck and cargo space for bulk supplies such as casing, tubulars, drilling fluid and cement. Standby boats form only a small part in the larger offshore emergency planning which is in operation (see Safety).

Safety: Offshore emergency plans are based on the principle of mutual aid between operators. This principle was formalised into "Sector Club" arrangements covering the main areas of operation in the UK Continental Shelf. For Offshore Shetland, UKOOA made arrangements with the Norwegian Industry Association for Operating Companies in 1978 to form the Orange Sector Club, operated by Elf Norway from 59 degrees to about 60 degrees 20 minutes North. From 60 degrees 20 minutes North, the other Sector offshore Shetland, the Red Sector Club, is operated by Shell UK. While each operator remains responsible for dealing with any emergency involving his own operations, as a member of a Sector Club he can call on the resources of the other operators in his Sector, or area to assist him.

Safety zones: By UK law, a safety zone with a radius of 300 metres is established around fixed offshore installations. No ship may enter this zone unless it is directly concerned with the operations of the installation. There are also safety zones for mobile drilling rigs and subsea wells.

Sweet gas: Hydrocarbon gas free from sulphur compounds.

Tanker-loadable crude oil: A crude oil which has a vapour pressure at a temperature of 100°F which is less than one atmosphere so permitting it to be shipped in simple non-pressurised tankers.

Tight hole: Any well for which the operating company is not prepared to announce the results.

Tonnage (ships). By international agreement, merchant vessels (eg. tankers, supply boats, and all such commercial ships mentioned in this book) are measured in tons of 100 cubic feet, the resultant tonnage being termed **Gross Tonnage**. Deductions are made from this figure in respect of engine-room and bunker spaces, accommodation and other non-commercial parts of the ship to arrive at the Net Registered tonnage. In the commercial employment of merchant vessels, particularly **tankers,** the tonnage usually referred to is the **Deadweight** (dwt) **tonnage**. This figure represents the total carrying capacity of the ship in tons of 2240 lb when loaded to summer marks (ie when loaded to the approximate capacity during the summer season). To arrive at the weight of the cargo carried, it is necessary to deduct bunkers, stores, water, etc. Whilst there is not necessarily any relationship between the Deadweight Tonnage, and the Gross or Net Tonnage, in the case of tankers the following relationship may be observed:— Approx. Summer Deadweight Tonnage = 2½ times the Net Register Tonnage. Gross Tonnage = approx. ⅔ of Deadweight Tonnage.

Tonne: One tonne = 2204 lb.

UKOOA: The UK Offshore Operators Association Limited, formed in 1973 from the informal UK North Sea Operators Committee which was established by the oil companies operating licences after the allocation of the first round of licences in 1964. Membership of UKOOA is restricted to the 34 companies presently acting as operators on production licences in UK waters. All oil companies engaged in exploration or production on the UK Continental Shelf are members. UKOOA is an oil industry forum for discussion of technical and administrative matters and this provides a means of communication with the Government and others. UKOOA has 18 permanent and two ad-hoc committees and is represented on 27 joint industry bodies. It appoints a Council to control its affairs annually and the Council appoints five Executive Officers. All members directly involved in production operations have an automatic right to be appointed to the Council and from the members engaged in exploration, five are elected and two may be appointed to the Council.

UKOOA have employed permanent staff since 1973, including a Director General, Executive Secretary and Technical Secretary.

UKOOA is concerned with Safety, amongst other matters, and a UKOOA representative is Chairman of the Shetland Aviation Committee which deals with all matters of concern to aircraft operators involved in the offshore oil industry, including air traffic control, and search and rescue systems.

Vapour pressure: The pressure exerted by vapour leaving the surface of a liquid.

Wildcat: Well drilled in search of a new oil or gas accumulation — an exploration well in a previously unexplored area or in a previously unexplored stratum.

Wireline: Any line of wire or cable used for downhole operations. Two types are usually distinguished; piano and electric wireline. The former is a thin single strand line of high tensile steel used to lower instruments or tools into a well, and/or to install, retrieve or operate 'wireline equipment' eg. fail-safe safety valves installed in tubing. Electric wirelines are normally used for surface recording instruments eg. those used for making electric logs. (See Chapter Three "Schlumberger Inland Services Inc.").

Workover: A re-entry into a completed well for modification or repair work.

Sources of Information: Bank of Scotland; BP; Daphne E. Duffy; UKOOA.

Sources of Further Information

1. Department of Energy, "Development of the Oil and Gas Resources of the United Kingdom", 1980, HMSO. Price £3.75.

2. Petroleum and Submarine Pipelines Act, 1975, HMSO. Price £4.50.

3. J. M. Raisman, "Oil and Gas — More to Come From Shetland Waters", Energy in the 90's Conference, Aviemore, 25th September, 1980.

4. Daphne E. Duffy, "North West European Continental Shelf Oil and Gas Field Development Survey", Institute of Petroleum, January, 1981.

5. Department of Energy, "A North Sea Gas Gathering System", Energy Paper 44, HMSO. Price £6.25.

6. Zetland County Council, "Interim County Development Plan", 28th March, 1973, (available from Shetland Islands Council, Department of Administration). Price £1.00.

7. Zetland County Council Act, 1974, Chapter VIII, HMSO. Price £0.95.

8. Zetland County Council, Director of Planning, "Sullom Voe District Plan", November 1974, (available from Shetland Islands Council, Department of Administration). Price £3.00.

9. Shetland Islands Council, Director of Planning, "Shetland Structure Plan — Report of Survey", April 1976. Vol. I, price £1.00; Vol. II, price £5.00.

10. James R. Nicolson, "Shetland and Oil", 1975. Price £4.75.

11. Shetland Islands Council, Research and Development Department, "Shetland in Statistics", No. 9, 1980, (obtainable from Shetland Islands Council, Research and Development Department, 93 St. Olaf Street, Lerwick). Price £0.75.

12. Civil Aviation Authority, "Sumburgh Airport Passenger Survey, 1977", June 1978.

13. Richard M. Suzman, MD; D. J. Vorhees-Rosen, RN, MA; D. H. Rosen, MD: "The Impact of the North Sea Development on Mental and Physical Health: A longitudinal study of the consequences of the economic boom and rapid social change", June 1980, New Shetlander.

14. Petroleum (Consolidation) Act, 1928, HMSO. Price £2.25.

15. SVEAG, "Oil Terminal At Sullom Voe — Environmental Advisory Group".

16. J. J. M. Gammack and M. G. Richardson, "A Compendium of Ecological and Physical Information on the Shetland Coastline", Nature Conservancy Council, June 1980.

17. The British Petroleum Company Limited, "Our Industry Petroleum".

18. Shetland Islands Council, Director of Ports and Harbours, "Port of Sullom Voe Information Book", 1981.

19. I. H. McNicoll and G. Walker, "The Shetland Economy 1976/77; Structure and Performance", Shetland Islands Council and Highlands and Islands Development Board, July 1978.

20. Local Government (Scotland) Act, 1973, HMSO. Price £1.80.

21. Social Work (Scotland) Act, 1968, HMSO. Price £1.85.

22. Manpower Services Commission, "Shetland Manpower Study", (in draft form), Shetland District Manpower Committee.